SO-CBI-338

# Science and Asian Spiritual Traditions

# Science and Asian Spiritual Traditions

GEOFFREY REDMOND

Greenwood Guides to Science and Religion
*Richard Olson, Series Editor*

**Greenwood Press**
Westport, Connecticut • London

**Library of Congress Cataloging-in-Publication Data**

Redmond, Geoffrey P.
Science and Asian spiritual traditions / Geoffrey Redmond.
    p. cm. — (Greenwood guides to science and religion)
    Includes bibliographical references and index.
    ISBN: 978–0–313–33462–7 (alk. paper)
    1. Religion and science—China.   2. Religion and science—India.   3. China—Religion.
4. India—Religion.   I. Title.
   BL240.3.R43   2008
   201′.65095—dc22        2007040227

British Library Cataloguing in Publication Data is available.

Library of Congress Catalog Card Number: 2007040227
ISBN: 978–0–313–33462–7

First published in 2008

Greenwood Press, 88 Post Road West, Westport, CT 06881
An imprint of Greenwood Publishing Group, Inc.
www.greenwood.com

Printed in the United States of America

The paper used in this book complies with the
Permanent Paper Standard issued by the National
Information Standards Organization (Z39.48–1984).

10 9 8 7 6 5 4 3 2 1

# Contents

Contents

# Series Foreword

For nearly 2,500 years, some conservative members of societies have expressed concern about the activities of those who sought to find a naturalistic explanation for natural phenomena. In 429 B.C.E., for example, the comic playwright, Aristophanes, parodied Socrates as someone who studied the phenomena of the atmosphere, turning the awe-inspiring thunder which had seemed to express the wrath of Zeus into nothing but the farting of the clouds. Such actions, Aristophanes argued, were blasphemous and would undermine all tradition, law, and custom. Among early Christian spokespersons there were some, such as Tertullian, who also criticized those who sought to understand the natural world on the grounds that they "persist in applying their studies to a vain purpose, since they indulge their curiosity on natural objects, which they ought rather [direct] to their Creator and Governor."[1]

In the twentieth century, though a general distrust of science persisted among some conservative groups, the most intense opposition was reserved for the theory of evolution by natural selection. Typical of extreme antievolution comments is the following opinion offered by Judge Braswell Deen of the Georgia Court of Appeals: "This monkey mythology of Darwin is the cause of permissiveness, promiscuity, pills, prophylactics, perversions, pregnancies, abortions, pornography, pollution, poisoning, and proliferation of crimes of all types."[2]

It can hardly be surprising that those committed to the study of natural phenomena responded to their denigrators in kind, accusing them of willful ignorance and of repressive behavior. Thus, when Galileo Galilei was warned against holding and teaching the Copernican system of astronomy as true, he wielded his brilliantly ironic pen and threw down the

gauntlet to religious authorities in an introductory letter "To the Discerning Reader" at the beginning of his great *Dialogue Concerning the Two Chief
World Systems*:

Several years ago there was published in Rome a salutary edict which, in order to
obviate the dangerous tendencies of our age, imposed a seasonable silence upon
the Pythagorean [and Copernican] opinion that the earth moves. There were those
who impudently asserted that this decree had its origin, not in judicious inquiry,
but in passion none too well informed. Complaints were to be heard that advisors
who were totally unskilled at astronomical observations ought not to clip the wings
of reflective intellects by means of rash prohibitions.

Upon hearing such carping insolence, my zeal could not be contained.[3]

No contemporary discerning reader could have missed Galileo's anger
and disdain for those he considered enemies of free scientific inquiry.

Even more bitter than Galileo was Thomas Henry Huxley, often known
as "Darwin's bulldog." In 1860, after a famous confrontation with the
Anglican Bishop Samuel Wilberforce, Huxley bemoaned the persecution
suffered by many natural philosophers, but then he reflected that the
scientists were exacting their revenge:

Extinguished theologians lie about the cradle of every science as the strangled
snakes beside that of Hercules; and history records that whenever science and
orthodoxy have been fairly opposed, the latter has been forced to retire from the
lists, bleeding and crushed, if not annihilated; scotched if not slain.[4]

The impression left, considering these colorful complaints from both
sides, is that science and religion must continually be at war with one
another. That view of the relation between science and religion was re-
inforced by Andrew Dickson White's *A History of the Warfare of Science
with Theology in Christendom*, which has seldom been out of print since it
was published as a two volume work in 1896. White's views have shaped
the lay understanding of science and religion interactions for more than a
century, but recent and more careful scholarship has shown that confronta-
tional stances do not represent the views of the overwhelming majority of
either scientific investigators or religious figures throughout history.

One response among those who have wished to deny that conflict constitutes the most frequent relationship between science and religion is to
claim that they cannot be in conflict because they address completely different human needs and therefore have nothing to do with one another.
This was the position of Immanuel Kant who insisted that the world of
natural phenomena, with its dependence on deterministic causality, is
fundamentally disjoint from the noumenal world of human choice and

morality, which constitutes the domain of religion. Much more recently, it was the position taken by Stephen Jay Gould in *Rocks of Ages: Science and Religion in the Fullness of Life*:

I ... do not understand why the two enterprises should experience any conflict. Science tries to document the factual character of the natural world and to develop theories that coordinate and explain these facts. Religion, on the other hand, operates in the equally important, but utterly different realm of human purposes, meanings, and values.[5]

In order to capture the disjunction between science and religion, Gould enunciates a principle of "Non-overlapping magisterial," which he identifies as "a principle of respectful noninterference."[6]

In spite of the intense desire of those who wish to isolate science and religion from one another in order to protect the autonomy of one, the other, or both, there are many reasons to believe that theirs is ultimately an impossible task. One of the central questions addressed by many religions is what is the relationship between members of the human community and the natural world. This question is a central question addressed in "Genesis," for example. Any attempt to relate human and natural existence depends heavily on the understanding of nature that exists within a culture. So where nature is studied through scientific methods, scientific knowledge is unavoidably incorporated into religious thought. The need to understand "Genesis" in terms of the dominant understandings of nature thus gave rise to a tradition of scientifically informed commentaries on the six days of creation which constituted a major genre of Christian literature from the early days of Christianity through the Renaissance.

It is also widely understood that in relatively simple cultures—even those of early urban centers—there is a low level of cultural specialization, so economic, religious, and knowledge-producing specialties are highly integrated. In Bronze Age Mesopotamia, for example, agricultural activities were governed both by knowledge of the physical conditions necessary for successful farming and by religious rituals associated with plowing, planting, irrigating, and harvesting. Thus religious practices and natural knowledge interacted in establishing the character and timing of farming activities.

Even in very complex industrial societies with high levels of specialization and division of labor, the various cultural specialties are never completely isolated from one another and they share many common values and assumptions. Given the linked nature of virtually all institutions in any culture it is the case that when either religious or scientific institutions change substantially, those changes are likely to produce pressures for change in the other. It was probably true, for example, that the attempts of

Pre-Socratic investigators of nature, with their emphasis on uniformities in the natural world and apparent examples of events systematically directed toward particular ends, made it difficult to sustain beliefs in the old Pantheon of human-like and fundamentally capricious Olympian gods. But it is equally true that the attempts to understand nature promoted a new notion of the divine—a notion that was both monotheistic and transcendent, rather than polytheistic and immanent—and a notion that focused on both justice and intellect rather than on power and passion. Thus early Greek natural philosophy undoubtedly played a role not simply in challenging, but also in transforming Greek religious sensibilities.

Transforming pressures do not always run from scientific to religious domains, moreover. During the Renaissance, there was a dramatic change among Christian intellectuals from one that focused on the contemplation of God's works to one that focused on the responsibility of the Christian for caring for his fellow humans. The active life of service to humankind, rather than the contemplative life of reflection on God's character and works, now became the Christian ideal for many. As a consequence of this new focus on the active life, Renaissance intellectuals turned away from the then dominant Aristotelian view of science, which saw the inability of theoretical sciences to change the world as a positive virtue. They replaced this understanding with a new view of natural knowledge, promoted in the writings of men such as Johann Andreae in Germany and Francis Bacon in England that viewed natural knowledge as significant only because it gave humankind the ability to manipulate the world to improve the quality of life. Natural knowledge would henceforth be prized by many because it conferred power over the natural world. Modern science thus took on a distinctly utilitarian shape at least in part in response to religious changes.

Neither the conflict model nor the claim of disjunction, then, accurately reflect the often intense and frequently supportive interactions between religious institutions, practices, ideas, and attitudes on the one hand, and scientific institutions, practices, ideas, and attitudes on the other. Without denying the existence of tensions, the primary goal of the volumes of this series is to explore the vast domain of mutually supportive and/or transformative interactions between scientific institutions, practices, and knowledge and religious institutions, practices, and beliefs. A second goal is to offer the opportunity to make comparisons across space, time, and cultural configuration. The series will cover the entire globe, most major faith traditions, hunter-gatherer societies in Africa and Oceania as well as advanced industrial societies in the West, and the span of time from classical antiquity to the present. Each volume will focus on a particular cultural tradition, a particular faith community, a particular time period, or a particular scientific domain, so that each reader can enter the fascinating story of science and religion interactions from a familiar perspective.

Furthermore, each volume will include not only a substantial narrative or interpretive core, but also a set of primary documents which will allow the reader to explore relevant evidence, an extensive annotated bibliography to lead the curious to reliable scholarship on the topic, and a chronology of events to help the reader keep track of the sequence of events involved and to relate them to major social and political occurrences.

So far I have used the words "science" and "religion" as if everyone knows and agrees about their meaning and as if they were equally appropriately applied across place and time. Neither of these assumptions is true. Science and religion are modern terms that reflect the way that we in the industrialized West organize our conceptual lives. Even in the modern West, what we mean by science and religion is likely to depend on our political orientation, our scholarly background, and the faith community that we belong to. Thus, for example, Marxists and Socialists tend to focus on the application of natural knowledge as the key element in defining science. According to the British Marxist scholar, Benjamin Farrington, "Science is the system of behavior by which man has acquired mastery of his environment. It has its origins in techniques . . . in various activities by which man keeps body and soul together. Its source is experience, its aims, practical, its *only* test, that it works."[7] Many of those who study natural knowledge in preindustrial societies are also primarily interested in knowledge as it is used and are relatively open regarding the kind of entities posited by the developers of culturally specific natural knowledge systems or "local sciences." Thus, in his *Zapotec Science: Farming and Food in the Northern Sierra of Oaxaca*, Roberto González insists that

Zapotec farmers . . . certainly practice science, as does any society whose members engage in subsistence activities. They hypothesize, they model problems, they experiment, they measure results, and they distribute knowledge among peers and to younger generations. But they typically proceed from markedly different premises—that is, from different conceptual bases—than their counterparts in industrialized societies.[8]

Among the "different premises" is the presumption of Zapotec scientists that unobservable spirit entities play a significant role in natural phenomena.

Those more committed to liberal pluralist society and to what anthropologists like González are inclined to identify as "cosmopolitan science," tend to focus on science as a source of objective or disinterested knowledge, disconnected from its uses. Moreover they generally reject the positing of unobservable entities that they characterize as "supernatural." Thus, in an *Amicus Curiae* brief filed in connection with the 1986 supreme court case which tested Louisiana's law requiring the teaching of creation science along with evolution, for example, seventy-two Nobel Laureates, seventeen state academies of science and seven other scientific organizations argued that

Science is devoted to formulating and testing naturalistic explanations for natural phenomena. It is a process for systematically collecting and recording data about the physical world, then categorizing and studying the collected data in an effort to infer the principles of nature that best explain the observed phenomena. Science is not equipped to evaluate supernatural explanations for our observations; without passing judgement on the truth or falsity of supernatural explanations, science leaves their consideration to the domain of religious faith.[9]

No reference whatsoever to uses appears in this definition. And its specific unwillingness to admit speculation regarding supernatural entities into science reflects a society in which cultural specialization has proceeded much farther than in the village farming communities of southern Mexico.

In a similar way, secular anthropologists and sociologists are inclined to define the key features of religion in a very different way than members of modern Christian faith communities. Anthropologists and sociologists focus on communal rituals and practices which accompany major collective and individual events—plowing, planting, harvesting, threshing, hunting, preparation for war (or peace), birth, the achievement of manhood or womanhood, marriage (in many cultures), childbirth, and death. Moreover, they tend to see the major consequence of religious practices as the intensification of social cohesion. Many Christians, on the other hand, view the primary goal of their religion as personal salvation, viewing society as at best a supportive structure and at worst, a distraction from their own private spiritual quest.

Thus, science and religion are far from uniformly understood. Moreover, they are modern Western constructs or categories whose applicability to the temporal and spatial "other" must always be justified and must always be understood as the imposition of modern ways of structuring institutions, behaviors, and beliefs on a context in which they could not have been categories understood by the actors involved. Nonetheless it does seem to us not simply permissible, but probably necessary to use these categories at the start of any attempt to understand how actors from other times and places interacted with the natural world and with their fellow humans. It may ultimately be possible for historians and anthropologists to understand the practices of persons distant in time and/or space in terms that those persons might use. But that process must begin by likening the actions of others to those that we understand from our own experience, even if the likenesses are inexact and in need of qualification.

The editors of this series have not imposed any particular definition of science or of religion on the authors, expecting that each author will develop either explicit or implicit definitions that are appropriate to their own scholarly approaches and to the topics that they have been assigned to cover.

Richard Olson

**NOTES**

1. Tertullian, 1896–1903. "Ad nationes," in Peter Holmes, trans., *The Anti-Nicene Fathers*, ed. Alexander Roberts and James Donaldson, vol. 3 (New York: Charles Scribner's Sons), p. 133.

2. Christopher Toumey, *God's Own Scientists: Creationists in a Secular World* (New Brunswick, N.J.: Rutgers University Press, 1994), p. 94.

3. Galileo Galilei, *Dialogue Concerning the Two Chief World Systems: Ptolemaic and Copernican* (Berkeley and Los Angeles: University of California Press, 1953), p. 5.

4. James R. Moore, *The Post-Darwinian Controversies: A Study of the Protestant Struggle to Come to Terms with Darwin in Great Britain and America, 1870–1900* (Cambridge: Cambridge University Press, 1979), p. 60.

5. Stephen Jay Gould, *Rocks of Ages: Science and Religion in the Fullness of Life* (New York: The Ballantine Publishing Group, 1999), p. 4.

6. Ibid., p. 5.

7. Benjamin Farrington, *Greek Science* (Baltimore: Penguin Books, 1953).

8. Roberto Gonzales, *Zapotec Science: Farming and Food in the Northern Sierra of Oaxaca* (Austin: University of Texas Press, 2001), p. 3.

9. *72 Nobel Laureates, 17 State Academies of Science and Seven Other Scientific Organizations. Amicus Curiae* Brief in support of Appelles Don Aguilard, et al. v. Edwin Edwards in his official capacity as Governor of Louisiana et al. (1986), p. 24.

# Acknowledgments

I especially want to thank my wife, Mingmei Yip, for her always gracious and inspiring presence and for her help with many aspects of this book, particularly translations of Chinese texts.

I am particularly grateful to my editor, Kevin Downing, for his patience and helpful advice. Also to the series editor, Richard Olson, for an inspiring dinner conversation that led to the work you now hold. Dr. Ruth-Inge Heinze, founder and president of Independent Scholars of Asia, has been a good friend for many years and first involved me in this project. I have benefited greatly from conversations with scholars of Asian science and medicine including Rahul Peter Das, Nathan Sivin, Paul Unschuld, Barbara Volkmar, Kenneth Zysk and many others too numerous to list.

The author and publisher gratefully acknowledge permission to excerpt material from the following sources:

Lynn, Richard John, trans. 1994. *The Classic of Changes: A New Translation of the I Ching as Interpreted by Wang Bi* (New York: Columbia University Press), pp. 142–48; 47–56. Copyright © 1994 Columbia University Press. Reprinted with permission of the Publisher.

Extracts from Mair, Victor H., Steinhardt, Nancy S., and Goldin, Paul R., eds. 2005. *Hawai'i Reader in Traditional Chinese Culture* (Honolulu: University of Hawai'i Press), pp. 24–26; 165–68; 440–43. Reprinted with permission.

Extracts from Xu, Gan. 2002. *Balanced Discourses: A Bilingual Edition* (New Haven and London: Yale University Press and Beijing: Foreign Languages Press), pp. 137–43; 175–85. © Yale University Press, 2002. Reprinted with permission.

Excerpts from Smith, Brian K. 1999. *Classifying the Universe: The Ancient Indian Varna System and the Origins of Caste* (Oxford: Oxford University Press), pp. 136–37. By permission of Oxford University Press, Inc.

Excerpts from Knoblock, John, and Riegel, Jeffrey, trans. 2000. *The Annals of Lu Buwei: A Complete Translation and Study*. Copyright © 2000 by the Board of Trustees of the Leland Stanford Jr. University. All rights reserved. Used with the permission of Stanford University Press, www.sup.org.

Extracts from Danielou, Alain. 1964. *The Gods of India: Hindu Polytheism*. Trans. from the French. Reprinted 1985. (New York: Inner Traditions International), pp. 28–34. Reprinted with permission.

Extracts from Needham, Joseph, with the collaboration of Ling, Wang. *Science and Civilization in China,* Vol.3. Mathematics and the Science of the Heavens and the Earth. © 1959: pp. 284–86; 367–69. Reprinted with the permission of Cambridge University Press.

Extracts from Unschuld, Paul. 1986. *Medicine in China: A History of Pharmaceutics* (Berkeley: The University of California Press). Reprinted with permission.

McKnight, Brian E., trans. 1981, Sung Ts'u *The Washing Away of Wrong. Science, Medicine, and Technology in East Asia 1* (Ann Arbor: Center for Chinese Studies, The University of Michigan), pp. 106–14.

Extracts from Furth, Charlotte, Zeitlin, Judith T., and Hsiung, Pingchen. 2007. *Thinking with Cases: Specialist Knowledge in Chinese Cultural History* (Honolulu: University of Hawai'i Press), pp. 154–55. Reprinted with permission.

Excerpts from Zysk, Kenneth. 1991. *Asceticism and Healing in Ancient India: Medicine in the Buddhist Monastery* (Oxford: Oxford University Press), pp. 22–25. By permission of Oxford University Press, Inc.

Lad, Vasant. 1984. *Ayurveda, The Science of Self-Healing: A Practical Guide*, pp. 100–103. Reproduced with permission from Lotus Press, a division of Lotus Brands, Inc., PO Box 325, Twin Lakes, WI 53181, USA, www.lotuspress.com. © 1984. All Rights Reserved.

# Chronology

What follows is necessarily selective, as a detailed chronology of two civilizations originating more than three thousand years ago would easily fill an entire volume.

## CHINA

Chinese history is divided into dynasties, each supposedly representing a single imperial family, although family relationships were sometimes invented to justify a powerful individual's accession to the throne. Events are often referred to by dynasty rather than exact date because each of the longest dynasties had a characteristic flavor in its spirituality, philosophy, science, art, and literature. The following chronology lists the most important dynasties, together with important events. For earlier dynasties dates are approximate.

**Mythical times and the Neolithic:** Many basic inventions such as the plow, fishing net, and medicine were attributed to mythical culture heroes such as Fu Xi, the sage kings Yao and Shun, and the Yellow Emperor (or lord). Because these inventions were made long before the development of writing, their origin is unknown. Archeologists have found bits of woven silk, pottery, carved jade, and other early technological products in Neolithic sites.

**Shang (1600–1045 B.C.E.):** The earliest body of Chinese writing, the oracle bones, record royal divinations and include some records of astronomical events. Bronze casting reached a high degree of technical and aesthetic perfection.

**Western Zhou:**  The earliest literary works, the Book of Songs and Yi Jing, were composed early in this dynasty.

**Warring States (475–221 B.C.E.):**  This was the second half of the Eastern Zhou and China's axial age, coinciding approximately with the era of the great Hellenic philosophers. It was the most fertile era of Chinese philosophy with lively debate between schools. Important thinkers of the Warring States include Confucius, Mencius, Xunzi, Laozi, and Zhuangzi. Later Chinese philosophy was mainly commentarial elaboration on the works of these seminal figures. Little is known about the lives of these figures and we cannot assume that the extant received texts were entirely authored by them. There are astronomical references in several of the classics, including the first recorded sighting of Halley's comet.

**Qin (221–206 B.C.E.):**  During the Qin dynasty, the first emperor, Qin Shi Huang Di, unified the separate states into the Middle Kingdom. His reign was despised by the later Confucians for its cruelty and for the emperor's order to burn all books without practical use—the *Yi Jing* was spared because divination was considered useful. Other canonical texts were later reconstructed from the memories of scholars. In his desire to keep the populace in a primitive, nonliterate state, the first emperor foreshadowed such twentieth-century dictators as Mao Zedong and Pol Pot of Cambodia.

**Former (Western) Han (206 B.C.E.–8 C.E.) and Latter (Eastern) Han (25–220 C.E.):** Chinese cosmology including the concepts of yin-yang and wu xing was elaborated into a complete system that endured until the twentieth century. Buddhism was introduced into China from India via the Silk Road.

**Northern Wei (386–584):**  Buddhism flourished and a unique style of sculpture was produced.

**Tang (618–907):**  Often considered the "golden age" of China. The Tang capital was cosmopolitan. Buddhism was at its peak. The armillary sphere was refined to permit more accurate observations. Several Indian astronomers held positions in the astronomical bureau. Buddhist monks brought Indian medical concepts and practices including cataract surgery but their impact on Chinese medicine was limited.

**Song (960–1279):**  Needham considered this the most fertile era for Chinese science. A notable scientist of the era was Shen Gua (1031–1095) who made notable observations of fossils and designed improved instruments for astronomical measurements (SCC I:1 134). Painting of nature reached its peak. Also during the Song the so-called Neoconfucian synthesis was created by Zhuxi, who incorporated elements of Buddhism and Daoism into Confucianism. This became a rigid orthodoxy that lasted until the twentieth century.

**Yuan (1271–1368):**  The dynasty of the Mongols who had conquered China. This was an era of relative religious tolerance and openness to foreign influences. Muslim influence on astronomy and medicine was prominent and Muslim Bureaus of Astronomy and Medicine were formed.

**Ming (1368–1644):**  The last dynasty ruled by native Chinese and the beginning of significant European influence. Matteo Ricci entered China as a Jesuit missionary in 1583 and remained there until his death in 1610. Ricci became fluent in Chinese and was accepted into court circles. He was particularly valued by the Chinese for his knowledge of mathematics and astronomy. Because of Church dogma Ricci employed the geocentric model, already being replaced in Europe.

**Qing (1636–1912):**  The emperors of the Qing were Manchurian, a fact much resented by many Han Chinese. Scientific activity increased as Western influence grew with the increasing presence of missionaries, traders, and officials from Europe. There was a trend toward more critical thinking in both science and philology.

**Republic (1912 to present. Continues only in Taiwan):**  The civil service examination system based on the ancient classics was abandoned in 1905. The May Fourth movement of 1919 advocated abandonment of much of the traditional ways of thought in favor of those of the modern Western, particularly science and democracy, as the only way forward for China.

**People's Republic of China (1949 to present):**  "Liberation," the victory of the Communists under Mao Zedong, was complete in 1949. Suspicion of anything that might be considered bourgeois, together with poor economic policy, stultified scientific activity. With the rise to power of Deng Xiaoping following the death of Mao, the economy progressively improved and Chinese scientists increasingly participate in international science.

## INDIA

The chronology of India is less easily summarized than that of China. While China in its bureaucratic style tended to keep detailed records with precise dates, this was not generally the case in India. Dating of early texts and events is often fanciful with a tendency to place them much earlier than modern scholarship supports. It has been suggested that India, because it conceived life as cyclical rather than linear, did not consider the events of a particular time of lasting significance. Other than stone inscriptions, few historical records survive from before the Mughal period. The dates of the various eras are approximations and sometimes overlap because what we now know as India was often subdivided into separate states.

**Indus Valley Civilization (Harappa Civilization) (3300 B.C.E. to 1900 B.C.E.):** The earliest known urban settlements in India. A variety of crops were cultivated. The major city of Harappa left seals with inscriptions but these remain to be deciphered. Harappa had some technology: houses were built of fired brick and a system of brick-lined drainage ditches is thought to be an early sewer system. As such this constitutes a very early public health measure.

**Vedic Era (1500–500 B.C.E.):** This is the era of Indian civilization associated with the formation of the *Vedas*, the foundational sacred texts of India. The last of these to be composed was the *Atharvaveda*, considered the origin of Ayurveda. As these works were orally transmitted for centuries before being set to writing, the dates of their origin are uncertain. Simple pottery and metalwork was made in this period. The Vedas contain many astronomical references and the profession of astronomer was recognized. Empirical observations were combined with cosmological speculations that persisted throughout premodern Indian history. Weaving of cotton was practiced in this era, before it developed in the West.

**Mahajanapadas (700–300 B.C.E.):** This consisted of a group of small states along the Indo-Gangetic plains. Hindu rituals were elaborate. Shakyamuni, the historical Buddha, lived and taught during this era, as did Mahavira, the founder of Jainism, the other major heterodox (non-Hindu) religion of India. Scholars do not question the actual existence of Shakyamuni but disagree on his exact dates.

**Achaemenid Empire (520–334 B.C.E.):** This Persian dynasty founded by Darius ruled Northwest India until the region was conquered by Alexander the Great. The influence of the Middle East is apparent in the use of Persian in Aramaic script. To a much greater degree than China, the distinctive civilization of India was the result of multiple foreign influences.

**Invasion of Alexander (327 B.C.E.):** The earliest event of Indian history that can be precisely dated. The conquest was complete by 334 and brought elements of Greek culture to India, including classical sculpture, as well as cosmological concepts. The contrapposto seen in Buddhist sculpture throughout Asia testifies to the persistence of Greek influence.

**Reign of Ashoka (273–232 B.C.E.):** Traditionally the greatest of all Indian rulers. In Buddhist tradition, it was Ashoka who promulgated the Dharma throughout India. The legend is that Ashoka did this as a result of remorse at the death toll of the Kalinga war. His seven *Pillar Edicts* placed throughout the empire contain moral instruction. One, strikingly for its time, calls for humane treatment of animals (Smith and Spear 1981:117–124).

**Kushan Empire (ca. 1st to 3rd centuries C.E.):** This was a state that ruled much of northern India and Central Asia. There was diplomatic contact with Rome, Sassanian Persia, and China. The Silk Road carried trade between East and the West. Buddhism has left many monuments in this region as it spread from India through to China. Caraka's seminal medical text, containing not only technical matters but also ethical ones, was composed during this era. The other foundational Ayurvedic text, that of Susruta, was written somewhat later but cannot be dated with any precision.

**Gupta Period (240–550 C.E.):** India's Golden Age. Stone sculpture reached a high point technically and aesthetically. This sculpture is said to have a pneumatic quality with the chest seeming capable of moving air, consistent with the central importance of breath in Indian yoga. Science began to develop during this era. A treatise on the gnomon, the *Pancasiddhantika*, written by Varahamihira in 505 summarized the astronomical knowledge of the time (Selin 1997:384).

**Journey to the West (399–414 C.E.):** The Chinese pilgrim Fa Xian traveled through India during this time in search of authentic Buddhist sacred texts. His written accounts are useful sources for Indian history of this era.

**Chola Empire (848–1279):** This empire stretched across Southeast India and traded extensively with Southeast Asia, bringing many aspects of Indian religion and culture, including not only Hindu iconography but also spicy food. Its technically accomplished cast bronze sculpture depicted gods, particularly Shiva and Parvati, in extremely sensuous fashion.

**Mughal Empire (1526–1707):** A time of great prosperity for India, under Moslem rulers. Akbar (1555–1606), one of India's greatest emperors, had a policy of treating Hinduism equally; this was later reversed under Aurangzeb (1659–1707), leading to revolt. Hindu orthodoxy became increasingly rigid, comparable to Neoconfucianism in China. The Taj Mahal was built under Emperor Shahjahan beginning in 1632. *Siddhanta*

texts incorporated elements of Moslem astronomy; observation and refinement of formulae by correction factors became highly sophisticated.

**Colonial Period (1757–1947):**  As a result of British colonization, Western
influence became substantial. In modern times India has distinguished
itself in science and medicine.

# Chapter 1

## Introduction: Science and Spirituality in Traditional Asia

Until rather recently, it was assumed that the function of a humanistic education was to provide the student with a deeper knowledge of "Western" culture, defined as beginning in ancient Greece, continuing through Rome, thence to Europe and to the United States and other countries settled by the British. That the study of cultures not included in this list might be of equal value was a consideration that rarely arose. Fortunately, the West is now moving beyond this rather narrow outlook. Many Asian cultural elements have diffused around the world from food to art, religion, and even, to some extent, language. Of great importance is the ready availability from mainstream publishers of translations of such Asian classics as the *Dao De Jing, Yi Jing, Bhagavad Gita,* and the *Tibetan Book of the Dead.* Furthering the trend is the ease of travel to previously remote places and the increasing role in the West of immigrants from countries such as India, China, Korea, and Vietnam.

When trying to understand and appreciate another culture, it is natural to compare it to one's own. Sometimes, the intent of such comparisons is to reaffirm the superiority of one's own culture; colonization was often justified in this way. An alternative trend, one that has now become prominent, is to search for solutions to our own problems in cultures whose ways are quite different. Beginning in the sixties, millions have made journeys to Asia, either physically or in their imaginations, motivated by disillusion with Western ways and the expectation that Asia had discovered ways for release from personal unhappiness that had eluded Westerners. While the number making these "journeys to the East" has multiplied, this search had its beginnings as early as the eighteenth century with philosophers such as Gottfried Leibniz who projected their utopian ideas onto a China they had never visited.

Though the China and India of Leibniz's time were to a great degree imaginary constructs, Western scholarship on these cultures has advanced greatly; now there are thousands of Western language scholars attaining a degree of erudition in Asian religious literature previously possessed by only a handful (Schwab 1984). Popularization has accompanied, and sometimes preceded, this explosion of scholarly interest. Yoga centers are everywhere and such terms as karma, nirvana, yin, yang, and Dao are now part of the English language. Tattoos with Chinese characters or Sanskrit mantras are popular with the young. Incense is burned for special occasions and decisions are made with the aid of the nearly three-thousand-year-old *Yi Jing*.

Unquestionably modern life has been enriched by the adaptation of these ancient traditions. Yet as often as not, the modern, popular versions of Asian spirituality coincide only approximately with their original Asian forms. Nor are the bits taken up in the West adequate samplings of the cultures of Asia. Popular accounts often give the impression that India and China offer spiritual depth, flavorful food, rising stock markets, and dense, impoverished populations, but not much else. Of particular importance, the fascination with Asian spirituality has obscured the scientific creativity of this part of the world. At most, some vaguely recall that the Chinese discovered gunpowder and the compass and that India discovered the notions of zero and infinity. This limited knowledge of the scientific accomplishments of traditional Asia is unfortunate, because science is not entirely a product of the past three centuries in the West but arose gradually in fits and starts in multiple locations. Indeed, until the last three centuries, science in China was at least as advanced as that of Europe.

There is now a considerable body of scholarship on Chinese science and somewhat less on India as well. Accurate understanding is impeded by the subject being covered only in specialist publications or else in popular treatments in which New Age ideas are substituted for accurate coverage. The inclusion of the present volume in the Greenwood Guides to Science and Religion indicates that the study of Asian science is now recognized as an integral part of the history of science.

## WAS THERE SCIENCE AND RELIGION IN TRADITIONAL ASIA?

There are special challenges in studying the history of science in Asia because there was no self-conscious category corresponding to the Western one of "science" or even the earlier European term "natural philosophy," which referred to deliberate but somewhat less systematic investigation of nature. Nor was religion or spirituality in general placed in a category distinct from other aspects of mental life. Thus we need to begin by considering what cultural elements in traditional China and India correspond to the contemporary categories of science and of spirituality. Adding to the challenge, we must consider intellectual developments over a period of more than three thousand years in two of the world's largest civilizations. No scholar can be familiar with more than a fraction of this material. In order to make its subject manageable, the present work is concerned only with traditional India and China, and only from the Bronze Age, when philosophy and curiosity about the universe began to flower, until the onset of predominant Western influence in the eighteenth century.

Science was always international to some degree but cross-cultural influence greatly accelerated beginning in the eighteenth century. Furthermore, such influences were for a long time absorbed into the existing culture rather than transforming it. In our time, the science of the West is the science of the entire world. Or, more accurately, the science that began in the West has become international science; papers from India and China are now published in English in the same journals as are research by Europeans or Americans. The institutional structure of science in universities, institutes, and industry in Asia is now similar to that of the West. While claims are occasionally made of different cultural styles in science, scientists are interested in the data and its implications, not in its cultural characteristics. True science is independent of its origins. (The notion that scientific truth is culturally determined has been briefly fashionable but was never taken seriously by scientists. While social factors influence science, for example in what projects get funded, they do not determine the results of experiments, though sometimes they transiently affect their interpretation.) There continues to be research on the traditional medicine of India and China but to the extent that this research has any influence on mainstream medicine, it uses the same empirical methodology as modern Western medicine. The term "Western science" is thus no longer accurate because this science is worldwide. Accordingly, when I use the term "Western science" it is comparison to the science of earlier times that was culturally specific.

Research on Asian science as a cultural and historical phenomenon has flourished in recent decades. Much of the fascination of Asian science is that its fundamental ideas have been recorded in ancient texts that are still extant. Both China and India show a striking continuity in their writing, mythology, cosmology, and underlying spiritual concepts. Although there were occasional book burnings, particularly that of the Qin dynasty of China, there was nothing comparable to the dark ages of Europe in which the links with classical culture were lost.

It would not be possible to offer a complete history of the development of all the major sciences in China and India in a work of this length; indeed, the basic research that would be needed for this is far from complete. Accordingly, I have adapted a more or less synchronic approach. That is, emphasis is on the nature of ideas rather than on their development over time. Salient aspects have been selected with the goal of making clear the general character of scientific, religious, and philosophical currents in both civilizations. This is in clear contrast to the evolutionary model that characterizes nearly all history of science in the West, in which science is seen as a progressively closer approach to the truth. Unavoidably, a scientist tends to see history of science in just this way because science does move toward progressively greater accuracy. Yet if we are to understand the past, it is not helpful to see it merely as an inferior version of the present. Yet it is best to acknowledge that this is not entirely avoidable, certainly with respect to science but also with regard to social structures such as use of torture for interrogation of suspects and the repression of women.

This presents particular difficulty for the study of science because modern scientific knowledge is far greater than that of traditional societies. The approach I have taken is to consider not only specific theories but also use of scientific method. My central theme is that empiricism as a means to approaching truth, though not explicitly formulated as a methodology in traditional Asia, continuously reemerges as humans seek to understand and control the natural environment. In this context, the ways truth is sought are as important as the theories and inventions that result.

Religion and metaphysical thinking generally were not so much antagonistic to science in India and China as intellectual structures that were not easily set aside because of their power in explaining the world. Indeed, the final abandonment of traditional metaphysics in favor of Western science was not the result of intellectual debate but because the superior effectiveness of Western technology—especially military—was sufficiently dramatic to reveal the limitations of the traditional ways of thought. We should not regard this, as did many Europeans who traveled in Asia, as due to backwardness but rather to the effectiveness of the traditional spiritual ideas in providing satisfying answers to religious, ethical, and metaphysical concerns. On the other hand, we need not out of sentimental attraction to such theories as yin and yang regard them as adequate alternatives to science.

## LIMITATIONS OF COMPARATIVE METHODOLOGY

Comparative approaches to cross-cultural scholarship have fallen out of favor, replaced by the concern to understand other cultures in their own terms as fully as possible. Though no one can be sure of completely setting aside tacit assumptions, we can aspire to become conscious of them. In the case of the history of science, however, comparison cannot be completely avoided since "science" as a distinctive form of knowledge is modern and Western. What we select as "science" in traditional China and India was not necessarily conceived of as distinctive forms of knowledge in those civilizations. Thus we are looking back to find aspects of thought that we, but not those who created it, categorize as science. The same is true of religion. No doubt, all realized that lighting incense to win the favor of a deity for an abundant harvest was a different activity than plowing a field, yet religion was not clearly demarcated as a special area of life. This demarcation is particularly strong in the modern West because of several factors that did not operate in Asia. First, in Europe, religious authority as vested in the Pope and church hierarchy was organized separately from state authority vested in a king, nobles, and counselors. Second, the rise in secularism in Europe was an effort to strengthen this separation. In China, religious organizations were more completely regulated by the state; when their wealth and power grew, emperors often repressed them with draconian measures such as destruction of monasteries and forced marriages of monks. Religious institutions did not enjoy the same degree of independence as Christian ones. There were of course ideological issues between state and religion, particularly celibacy which was contrary to Confucian teachings about the responsibility to maintain the male family line—only male children could offer the appropriate sacrifices to a person after his death. State and religious establishments at times conflicted over other issues because the state never considered religion as autonomous.

## CONSIDERING WHAT IS ASIA?

Asia is both the largest of earth's continents and an area of distinctive cultures. Though Asian cultural elements such as science, religion, philosophy, and art are commonly grouped together, we must keep in mind that all of these aspects of culture show extraordinary diversity. Recent trends in scholarship decry "essentialism," the assumption that cultures have a fixed essence in all their expressions, for example that there is something inherently "Chinese" in all aspects of that country's culture. This issue is easily avoided simply by sidestepping it. We can say that much of the religion, philosophy, and science produced in China have certain traits that mark them as Chinese simply as historical fact, without implication that these traits constitute an essence of Chineseness. Similar is the case for India. However, we should not allow fear of the bogeyman of essentialism deter us from seeking unique aspects of different cultures.

This is as good a place as any to justify frequent use of "the West" as a specific geographical and cultural area. There are many problems with this usage, despite which the term remains unavoidable. First, the assumption of a direct line from Greece to modern Europe and America is a retrospective selection that fails to consider the influences of other cultures of areas such as India, Persia, and Arabia. In the present time, much of what we think of as Western is now worldwide. A detailed critique of what constitutes the Western cultural tradition is beyond the subject of this book. Here I only wish to note that the concept of an entirely culturally distinct part of the world called the West is far from unproblematic. Nonetheless, the term is unavoidable. I must also clarify what I mean when I refer to "our culture" in contrast to India and China. I do not mean to restrict this to any particular geographic area, but simply to contrast the modern cultures in which we are all immersed with that of traditional China and India.

## WHY INDIA AND CHINA?

The Asian continent is now divided into more than fifteen distinct countries, as culturally distinct from each other as the cultures of European countries. For practical reasons of length, treatment is narrowed to India and China. However, it is misleading to assume that the cultures of these countries were ever coextensive with their present or past national boundaries. Rather they were the largest centers of the two most widespread cultures of the Asian continent. Many common cultural elements are shared by China, Vietnam, and Korea, for example. Thus Korea considers itself more Confucian than China, and even uses the *bagua* (*Yi Jing* trigrams arranged in a hexagon) as its national symbol. However, it is not quite correct to say that Korea's culture is derived from that of China. Rather, both have common cultural elements that have developed in distinctive forms. To describe this as China influencing the cultures of these two smaller countries is rather misleading. A more accurate understanding is that all shared common cultural elements, for example a writing system and the cluster of ideas termed Confucianism.

When I present China and India as influencing the other cultures of Asia, I am not implying that the role of other Asian countries was simply passive absorption. Rather, all participated in the creation of these rich cultures. Similarly, diverse countries from Sri Lanka to Nepal share in what we tend to lump together as Indian culture. It is more useful to divide Asia into two large cultural zones: South Asia, consisting of those whose culture is what we more narrowly call Indian, and East Asia, consisting of China, Korea, Japan, and Vietnam. What is now referred to as Southeast Asia was formerly termed Indo-China, a term that is more geographic than cultural.

I will not be concerned to identify the exact geographical origin of each idea; often such matters are highly contested for reasons of national pride in the contemporary world.

## THE LANGUAGES OF INDIA AND CHINA

One extreme difference between India and China was—and is—language. Sanskrit, the language of India's elite for most of its history is an agglutinating, that is, highly inflected, language in which the grammatical role of a word is evident from its endings. In this, Sanskrit is similar to Latin but more elaborate. Writing is phonetic, making pronunciation apparent from the writing. Chinese is an isolating language; it has no prefixes or suffixes to indicate the role of each word in a sentence. Rather, each word's function is indicated only by its position. One theory is that Chinese originally had inflections but that these were lost because the Chinese writing system, being nonphonetic, cannot record them. Chinese is written in abstract symbols, one or more per word, a system that does not permit inflections to be recorded. The benefit of this writing system being largely independent of sound is that Chinese people speaking different dialects can read the same text and understand the meaning, even though their spoken languages are mutually incomprehensible. That the same written language could be employed all over China is an important reason for the cultural homogeneity that characterized China despite its immense population and broad geographical spread.

## THE PLAN OF THIS BOOK

Our inquiry will begin by considering what distinguishes science as a specific form of knowledge and what sorts of ideas and practices can be categorized as spiritual. In place of judgmental categories such as pseudoscience and superstition, a central distinction between metaphysical and empirical thought is utilities to define which ideas are scientific and which are not. Metaphysics works on the basis of thought alone and may be entirely rational or may involve the supernatural. Empirical thinking, which is the basis of science, involves confirmation of hypotheses by actual testing. Hypotheses do not become part of science until they are validated by systematic observation or by experiment. Science avoids directing its efforts toward matters that cannot be resolved by systematic data gathering; though it may speculate based on existing data, speculations eventually are accepted or rejected on the basis of empirical test.

Correlative thought is the way traditional cultures organized the confusing diversity of the world, a way that is systematic but in a quite different way than science. Indeed, correlative thought explains the world with sufficient plausibility as to resist replacement by science. Correlative thought is at root metaphysical.

For the sake of coherence, I have followed a set order in which I consider China first, then India. In general, my treatment of China is more extensive. There are several reasons for this. First, English language scholarship is far more complete regarding Chinese science than Indian science. This is due in no small part to the accomplishments of one man, Joseph Needham, whose monumental *Science and Civilization in China* not only assembles immense amounts of information but also inspired many historians in the direction of Chinese science. Even for those who disagree with Needham, his work provides the necessary background for beginning investigation. Unfortunately, we do not have any comparable work for Indian science, though a new series is promising (Rahman 1998).

Another reason for my emphasis on China is that the thought of India has greater continuity with that of the West, both in terms of the language and the preference for abstract formulations. Being adjacent to Persia and having been invaded by Alexander, India was much more influenced by the same strains of thought as the West, though it developed them in quite different ways. For example, Indian metaphysics was far more abstract than that of pre-Buddhist China and thus is closer to the style of thought of scientific physics than is the metaphysics of China. I thus present first the system less like our own and then go on to one that may be regarded as intermediate. This is simply a convenience to make it possible to cover a wide range of ideas between cultures.

Many other subjects and approaches could have been chosen. My hope is simply that this book will serve to awaken interest in both science and spirituality in these two ancient Asian cultures and to give something of the flavor of the thought of each. The subjects are vast; the annotated bibliography will serve to point the way for anyone wishing to pursue any of the subjects reviewed in the present work in greater depth.

## METHODOLOGY AND ASSUMPTIONS

My approach in this book is to emphasize scientific *method* as much as the *content* of scientific knowledge. This means that some of the subjects I cover, such as ceramic manufacture, belong to what tends to be classified as technology, tacitly assumed by many scholars to be a lesser sort of knowledge because its intent is not discovery of general principles. From my perspective, much of the technology of traditional cultures is actually closer to science than, for example, speculative cosmology, because technology must involve empirical testing of some sort.

My emphasis on development of scientific method rather than solely on its content is due to my background as a biomedical scientist while that of most historians of science is in the physical sciences or mathematics. Historians of science tend to emphasize the search for basic laws of nature, such as Newton's laws of motion, the ideal gas laws, or Einstein's special theory of relativity. Popular books on science also take these to be the final goal of science. Even Joseph Needham, though a biochemist, seems to have thought more as a chemist than a biologist. Such assumptions give little idea of what working scientists do every day. For us, the laws of nature are simply the principles we assume in making other discoveries. Thus the discovery of the double-helical structure of DNA, one of the most fundamental advances in human knowledge, is now mainly thought of as the basis for making further discoveries, such as sequencing the human genome. Especially in medical research, the goal is practical: the cure of disease. Medical research values empirical findings over search for fundamental principles. With new treatments invented, subsequent research is concerned less with the scientific laws involved than with whether it works and is safe. The phenomena of biology are too diverse to be reduced to a few fundamental laws; in this respect it is quite unlike physics. This means that for biology, the most fundamental principle is the method of empirical verification. Speculation is viewed as self-indulgent. Theories do not advance biomedical science, experimental findings do.

A fundamental distinction in my approach, to be developed at greater length throughout this book, is between empirical and metaphysical thinking. Metaphysics depends on reasoning alone while empirical method involves some actual testing. I do not suggest that these two modes of knowledge can be rigidly separated. Thus in China, astronomers made systematic observations of celestial events which might then be metaphysically interpreted to be heaven's judgment on the Emperor's governance. The blend of empirical and metaphysical can also be seen in smallpox vaccination, which was shown effective by actual observation but explained metaphysically. Metaphysics however is never science and generates accurate predictions about the world only by accident. It can explain everything after the fact; this is how it survives empirical refutation. Failed prophecies can always be rationalized as incorrect interpretation of the oracle.

I do not want to suggest that metaphysics is an inferior mode of thought. Nonempirical speculation about the universe and our place in it is universal. For many, it responds to intellectual and spiritual concerns that science does not. I do however hold that purely speculative thought must be distinguished from science. Though I may seem at times in what follows to use the term "metaphysical" as derogatory, that is not my intent. Certain fundamental human questions, such as What is the purpose of life? How can I cope with suffering? What is the proper way to live? What happens after death? can only be addressed by metaphysical reasoning. However, other questions such as how is a particular disease to be cured or what foods are healthiest clearly are more effectively resolved by empirical test.

My critique of metaphysics refers to situations in which it is employed to address empirical questions. In the traditional world, metaphysics was used to resolve fundamental questions because the nature of empirical methodology had not been clearly recognized. We find the inverse of this in the modern world when science is misapplied to dismiss religious beliefs. I definitely do not hold the view that science by itself gives a complete account of human life. However, for understanding physical and biological phenomena, including human disease, denying the efficacy of science is simply perverse. Yet, metaphysical thinking is inherent in the human mind and there seems no point in pretending that we can do without it. Such ideas as freedom, democracy, human rights, and others are essential for humanity, despite not being empirically verifiable.

My assumption is that the value of both forms of knowledge is enhanced when their differences are understood. Thus, we no longer fear eclipses because we understand that they result from particular positions of the sun and moon and have no ominous significance. Similarly, knowing that diseases are not caused by curses lessens the chance of innocents being condemned as witches. Of course science too can be misapplied for evil purposes, as in the eugenic experiments of the nineteenth and early twentieth centuries. Such abuses were not in fact based on real science, only on claims of being scientific. It is important that such abuses be recognized in the hope that they will not be repeated. However, I mention them here simply to illustrate that a simplistic notion of science as always benevolent but metaphysics as fallacious is not valid.

## THE SUPPOSED ANTAGONISM OF SCIENCE AND RELIGION

The history of science has tended to be presented in triumphalist terms as truth winning out over ignorance and superstition. In particular, sensational events such as the trial of Galileo present science as in battle with religion. Though not without a grain of truth, this is a very limited paradigm by which to understand the relation of religion and science. (For a useful account, demonstrating the degree to which protestant orthodoxy was a motivation for study of the natural world, see Olson 2004.) Particularly in Asia, science was not often in overt conflict with religious institutions or metaphysical ways of thought. It was however controlled by the state, which always feared knowledge that it could not control. While religion, by offering a comprehensive view of the universe, may have inhibited science by satisfying curiosity, it was not in opposition to science. Indeed any science-religion duality we find in traditional Asia is imposed by us.

To see science as progressively achieving more complete and accurate knowledge is a retrospective view with the errors and blind alleys to be de-emphasized. This evolutionary model is particularly unsuitable for characterizing the science of traditional Asia, which often did not show systematic progress toward more complete understanding. Rather, isolated discoveries were made but usually were not integrated into a systematic understanding. Not infrequently, what seem in retrospect to have been significant discoveries were simply forgotten. More useful is to consider the science of premodern China and India as recurring manifestations of the human desire to understand the world and intuitive application of the principle of empirical verification.

With our current awareness of the sad effects of colonization and discrimination, discussion of cultural differences makes some uncomfortable. Yet they account for much of the dynamism of human existence. I admit to being fascinated by these differences but still need to state explicitly that recognizing differences is not to imply any judgment of superiority or inferiority. That I take up China first in each section and sometimes discuss it at greater length is simply due to my own interests and knowledge and not at all a judgment that one of these two cultures is somehow more important. Nor is there any suggestion that the science of one of these cultures was superior to the other. Both made profound discoveries. Both displayed extraordinary spiritual creativity, producing ideas and practices that have spread around the entire world.

In selecting which sciences to cover, I have chosen areas related not only to my own interests but also to those of greatest contemporary relevance. Thus medicine and ecology receive extended treatment. Since this work would be incomplete without inclusion of physical sciences, I have devoted a chapter to astronomy, arguably the oldest science. That the heavens changed in regular patterns greatly influenced the human speculation about the universe. Then, as now, the sky was the locus of the most important spiritual entities. Observation of the regular cycles of the seasons and heavenly bodies is probably the origin of the universal human assumption that a coherent pattern underlies human existence.

A considerable body of scholarship exists regarding Chinese and Indian alchemy but I have omitted coverage of this area because my own view, contrary to that of Joseph Needham and some other historians of science, is that alchemy has minimal relation to scientific chemistry. While alchemy does have a spiritual component, this has been much exaggerated following C. G. Jung's influential, but dubious, reinterpretation of it as a vehicle for self-transformation. Alchemy both East and West seems to me more of a curiosity in the study of human error than something of general interest. A useful critical analysis of mistaken interpretations of European alchemy as primarily spiritual is Principe and Newman (2001).

## NOTE TO THE READER

Chinese has been transliterated in the modern pinyin system. Sanskrit terms are rendered without diacritical marks. In many cases the Chinese and Sanskrit terms have been English words. They are italicized at first appearance in a chapter, but not thereafter.

Unattributed translations from the Chinese are by Mingmei Yip and myself.

# Chapter 2

—~—

# Understanding the Nature of Science and Spirituality in Other Cultures

If we are to seek to understand premodern cultures, we must set aside the assumption that they are simply variants of our own. Neither China nor India, the two Asian civilizations considered in the present work, had explicit concepts equivalent to our own categories of science and spirituality. These are notions we apply to those countries in retrospect. We cannot proceed, therefore, until we have considered what aspects in these cultures we can classify as science and as spirituality. This is a daunting task as we shall be considering a total of more than six thousand years of history in two of the creative and most populous regions of the world.

Within our own culture, to define science and religion, we can start by looking at the institutions that embody them. Thus physics is what is taught in physics departments and religion is what goes on in churches, synagogues, and temples, as well as university departments of religious studies. This approach, however, does not tell us why certain subjects came to be placed in their respective categories. Nor are these categories always applied carefully. Given the prestige of science, theories are often claimed to be scientific when they actually are not. Alternative medical practices are an example. Many remedies sold on the Internet, for example, are advertised as "scientifically proven" when there is no real research demonstrating effectiveness. Similarly, organizations may claim to be religious simply to obtain tax-free status. Others, such as UFO cults, may deny this label and claim to be scientific. So the inductive approach to defining science and religion is not of much use in defining the margins of these categories. Most important for the present inquiry, institutional structures comparable to our universities had not formed in Asia during the time period we are covering, so that this sort of definition does not take us far. Nor can we simply tell ourselves that we know science and religion when we see them. More precise definitions are needed to define the fields for study. Some aspects, particularly regarding the nature of science are contested by modern scholars, but since our focus is on historical matters, we shall try to simply step around controversies that have no direct bearing on our main focus.

## WHAT MAKES KNOWLEDGE SCIENCE?

Science is a body of knowledge about such matters as the structure of the universe, the nature of matter, the transmission of traits by genes and the causes of disease. The subjects and methods of science are clear enough in the modern world but less so in traditional cultures. We can identify the main defining attributes of science upon which virtually all scientists, and most philosophers, would agree:

1) Science takes *the working of external world* as its subject, e.g., cosmos, body, nature. Even when the workings of the mind are studied, they are measured by external, objective means. Thus emotions might be studied by means of numerical rating scales or even as changes of blood flow in specific brain regions.

2) Scientific knowledge is systematic. It collects facts and generates hypotheses to organize them into a coherent pattern.

3) Science is *cumulative*; it is continuously building upon and modifying earlier knowledge. New knowledge both depends on the old and supplants it. Science locates its truths in the present and looks to the future for ever more accurate knowledge. Religion, in contrast, tends to seek truth in its past, from sacred texts and traditions considered to have greater authority over more recent ones. Though innovation is recurrent in religion, as in other areas of culture, it tends to be understood by believers as a return to original truths.

4) Scientific theories must be *verifiable*, that is capable of being tested by objective means. Thus we postulate the existence of a force called gravity because all bodies are attracted to each other just as a stone is attracted by the earth and so falls to it. Strictly speaking, "gravity" does not refer to anything tangible, but to a characteristic way all objects behave.

The principle of verifiability has been contested on a number of grounds but remains how scientists define their form of knowledge. The twentieth-century philosopher of science Karl Popper challenged the principle of verifiability, pointing out that theories can never be proven, only falsified. Thus the fact that the sun has risen every morning does not absolutely prove it will rise tomorrow. For Popper, scientific theories are never absolutely verifiable. To some degree this is a verbal quibble. For verifiability

Popper simply substituted the principle of falsifiability. A scientific proposition must be capable of refutation in the face of additional experience. Thus gravity could be falsified if an exception were found, however implausible this would be. On the other hand, that God exists, cannot be falsified because no empirical test can be imagined that would prove this. Thus universal gravitation is a scientific proposition while the existence of God is not. Note that this does not mean that God does not exist, even though some take it that way. A more nuanced interpretation is that belief in God is of a different sort than belief in gravity. In the indigenous Asian religions examples of nonfalsifiable concepts include not only the existence of a great variety of deities, but also rebirth, the existence of an underlying principle termed the Dao or Atman, and of a state beyond suffering termed nirvana, to name just a few.

The British empiricist philosopher A. J. Ayer declared any proposition for which we cannot devise a means of empirical verification to be "unintelligible." This is a form of "scientism," the belief that only experimentally verifiable propositions constitute authentic knowledge. It ignores the obvious fact that all of us hold nonfalsifiable beliefs that are essential to our lives, ethical ones as well as religious ones. We need not abandon the criterion of verifiability, rather, we can distinguish between scientific and other forms of human knowledge or belief. This is essential when we consider the scientific ideas of premodern cultures, in which the notion of empirical verification was never explicit and in which scientific and other forms of knowledge were mixed together.

One distinction, which will be central to what follows, is between scientific and metaphysical thought. While a scientific statement can be disproved, metaphysical propositions, because they deal with what is beyond direct experience, can neither be empirically verified nor falsified. Thus the theory that the interaction of yin and yang underlies all phenomena is metaphysical because there no experiment can be devised that will test for the presence of yin and yang. Metaphysical statements are the product of pure thought and may appear to have great explanatory power, but they are not science. Some metaphysical propositions are more coherent and plausible than others, of course, and debate about metaphysical matters has been a major preoccupation of philosophy since its beginnings.

Most religious propositions, such as the existence of God or other supernatural beings, are metaphysical. There are, of course statements *about* religion that are verifiable in principle, such as the dates of the Buddha's life—though evidence may be inadequate for a firm conclusion. There are other religious statements that are not verifiable now but might be in the future. No objective evidence has demonstrated that there is life after death but it is conceivable that such evidence might one day be discovered, though it is hard to imagine how what form such evidence would take. Like other scientists, I accept verifiability (or falsifiability) as the basis of scientific knowledge, but to label metaphysical propositions as "unintelligible" does not follow from this. Such statements as, "Your illness is due to an excess of kidney yang," or, "I believe in God" or "I am earning merit in hope of a favorable rebirth" are quite intelligible. They are just not objectively verifiable. Much of the study of religion is concerned with the human meaning and significance of unverifiable propositions.

In practice, an absolute distinction between empirical and metaphysical thought is not always possible or desirable. Meditation, for example, produces changes in blood pressure, brain waves, and hormone levels that have been objectively verified. However, some of the skills claimed by experienced meditators—and I include myself in this group—such as sensing the flow of the subjective forms of energy termed *qi* in China and *prana* cannot be measured. Yet as inner experience they are real. Meditation, like other religious practices, is most adequately understood as functioning at once on objective, subjective, and metaphysical levels. This is the sort of approach the present book tries to achieve: considering the different sorts of knowledge together, distinguishing them clearly but not dismissing either.

Lloyd and Sivin, like most historians of science, define the science of traditional cultures by its content rather than its methods:

Yet, in the ancient world, people already entertained ideas about the stars, the human body, the variety of living beings, the composition of things and the changes they undergo. Inquirers attempted . . . to understand the world that lay outside the social realm. . . . We will use "science" . . . to cover such studies as these. The mark of science, in that usage, lies in the aims of the investigation and its subject matter . . . not in the degree to which either the methods or results tally with those . . . of modern science. (Lloyd and Sivin 2002:4)

This exemplifies the laudable efforts of contemporary scholars to consider earlier civilizations in their own terms, rather than as inferior versions of our own. The emphasis of Lloyd and Sivin is with the content of science, rather than its method. In the present book, however, I give considerable attention to method. I do cover specific aspects of the content of the science of premodern Asia, but I am particularly interested in the extent to which scientific method was recognized and applied as a special way to knowledge. As we shall see, while an explicit philosophy of verification did not develop in traditional Asia—or in the West before the eighteenth century—when the method was applied it often met with great success.

This focus on method is essential when science is considered in relation to spirituality because in order to do so, we inevitably distinguish empirical from spiritual or metaphysical knowledge. Here we are less interested in how much Chinese and Indians knew about astronomy, for example, than in how they thought about the universe around them.

The role of social factors in determining what science holds to be true is a question much emphasized in recent work in the philosophy of science. That science is influenced by social factors is less shocking than it is sometimes made to seem. No scientist would deny the effects of social factors on the progress of science. New findings are too often ignored or derided, only to be accepted much later. Experiments are not always flawlessly conducted; even when properly done they may be misinterpreted. Science, like spirituality, is also an ideal not always fully attained. Nonetheless, the criteria I have given do accurately describe what science aspires to be.

When interpreting the science of traditional cultures, a number of interests are involved which may not be explicitly stated. The tendency to present non-Western cultures in a negative light, termed "Orientalist" has been much criticized as covertly intended to justify colonialism. The present work assumes that the achievements of Chinese and Indian traditional cultures are best appreciated by attempting the most accurate possible historical reconstruction. That science reached its fullest development in Western culture is unarguable. Yet China made many important inventions and discoveries, such as the paper, which makes this book possible. India developed observational astronomy to a high degree of accuracy. Yet the predominant mode of intellectual analysis in both civilizations was correlative rather than causal.

## THE PRACTICE OF SCIENCE: THE CENTRAL ROLE OF EMPIRICAL VERIFICATION

While thinking up theories seems glamorous and tends to be emphasized in elementary science teaching, most of scientists' days are spent tediously gathering and analyzing quantitative data. The quest for data unites scientists despite differences in age, gender, social class, culture, religion, academic discipline, politics, and everything else. For every scientist controlled observation is where valid knowledge begins. Speculation, while pleasurable and a large part of popular science articles, is discouraged. Thus, during a recent friendly debate about an issue in hormone therapy, a scientist-friend remarked in response to some theorizing on my part, "In God we trust, all others show need data."

I am emphasizing this point because in my view as a research scientist, historians of science, and popular writers as well, have tended to assume that the purpose of science is to formulate general laws, such as conservation of energy and mass or Newton's laws of motion. It is more interesting for most to read about how Newton discovered his laws than to try to understand the equations. For the working scientist such laws are not goals but tools. Most of us do not aspire to discover fundamental laws of nature, but simply hope to gather little bits of information that help complete a much larger pattern. For a brilliant few, these little bits are important enough to be awarded Nobel prizes. All this is by way of explanation as to why my background as a working research scientist leads me to emphasize the role of empiricism in the science of the past.

The emphasis on general laws is in large part due to taking physics as the model or "ideal" science. However, biomedical science is just as real and at least as important. While it is often assumed that all science can be derived from the laws of physics, this is not true in practice now and is not likely to be in the foreseeable future. Biological and other complex systems must obey the laws of physics but are not entirely predicted by them. While there are fundamental discoveries that underlie much of biology, such as the already mentioned double-helical structure of DNA, the goal of biology is as much to find the implications of such underlying principles as to discover the principles themselves. To give but one example, knowing the chemical composition of a particular gene does not enable one to know the function it codes for, nor whether it is a normal or disease-producing gene. These matters, of vital importance for both basic understanding and for medical therapeutics, can only be worked out by further research.

Now these points must not be pushed too far. Most of what each of us knows is acquired without putting it to systematic empirical test. We can be quite certain of much of our own subjective knowledge, such as what foods we like. Much of what we know about the natural world is acquired by casual observation—that plants cannot survive without water, for example—that hardly needs any more systematic test.

While systematic empiricism developed only sporadically in traditional Asia, this does not mean that observation was deficient. There is no doubt that ancients knew the configurations of stars and planets far better than moderns, only excepting amateur or professional astronomers. Divination by astrology or by the study of the cracks induced by applying heat to bovine scapulae or tortoise plastrons required careful observation of the resulting patterns. Professional diviners became experts on these patterns. Animal portrayals in Chinese painting were extremely accurate and an essential source for Western natural history illustration, including that of James Audubon. In Indian sculpture, the dwarf of ignorance who is portrayed as being stomped on by Shiva is a quite accurate depiction of the medical condition of achondroplasia. The ancients were no less able to observe with care than we are. What is different is the system that the observations were fitted into. Classificatory schemes are necessary for data to be understood and remembered. In earlier India and China, these were based on mental correlations, not objective study. These correlative schemes had spiritual significance because they described an orderly world that functioned by comprehensible principles such as yin-yang or the three gunas. These principles were in a sense more human than, for example, Newton's laws of motion, because emotion was inherent in them. Thus yang, or sattva, represented organization, positive development, while yin or tamas, represented deterioration, loss of vigor and so on. (The meaning of these cosmological principles is discussed in more detail in subsequent chapters.) Even now, scientific terms are used metaphorically. When we say a thought-pattern is right or left brain, or that we feel inertia; or that someone attractive is "hot," correlative thinking is occurring.

## PROTOSCIENCE, PSEUDOSCIENCE, AND SUPERSTITION

Deciding what early forms of thought can be considered science is far from straightforward. Terminology is a delicate matter because labeling something as "unscientific" seems pejorative. Hence, we need to reconsider some terms commonly used in history of Asian science. "Protoscience" has been used to refer to early ways of thought that have some attributes of science. Protoscience is assumed to differ from "pseudoscience," which mimics the form of science but not its method. This distinction can be problematic. For example, the search for substances or practices to confer immortality, a common undertaking in China, can be considered to be a predecessor of empirical medicine, or as a mere pseudoscience based on correlative metaphysics. While scientific medicine bases its methods of life extension on large-scale clinical trials, such as the studies on tens of thousands of men and women showing that lowering cholesterol reduces the risk of heart attack, nothing like this was done in China or India, nor was there anything resembling scientific conceptions of what might prolong life. Indeed, some immortality elixirs were rapidly fatal. Modern ways of increasing life expectancy did not develop in continuity from older ones. Rather they required a shift from a metaphysical to an empirical understanding of life processes. Hence it would seem more accurate to label alchemy as pseudoscience. Yet whether we consider alchemy a protoscience or a pseudoscience does not add much to our understanding, certainly it brings us no closer to understanding what the Daoist savants of alchemy thought they were doing.

In the present work, for the old protoscience/pseudoscience discourse, I have generally substituted an analysis emphasizing the distinction between metaphysical and empirical approaches. This avoids the need to make retrospective value judgments without losing sight of when earlier ideas were scientific and when they were not. Often, as we shall see in the case of astronomy and astrology, pseudoscience and protoscience cannot be separated. Nonetheless, with some traditional subjects, such as feng shui, though complex calculations were often involved, it is to resist the label of pseudoscience since no empirical basis for its theories are evident. Performance of calculations and use of a compass with mysterious markings no doubt made feng shui more impressive to clients but that did not make its pronouncements at all scientific.

Another category that needs to be carefully considered when looking at beliefs of the past is superstition. This is an even more pejorative term than pseudoscience and is usually applied with polemical intent. Labeling beliefs and practices as superstitious can come from seemingly opposite directions. Those with a materialistic, scientific outlook may call any belief superstitious that is contrary to established scientific orthodoxy as, for example, astrology. However from the side of religion, beliefs unacceptable to a particular orthodoxy are often condemned as superstitious. This term appears frequently in the accounts of European missionaries who were shocked by the native beliefs and rituals. Particularly offensive to Christians was "idolatry," the supposed worship of images, always integral to the non-Moslem religions of Asia. This prejudice stems from the prohibition against "graven images" in the Ten Commandments and is strongly inculcated in Western culture. Yet outside the mental realm of the religions of the Book, ritual use of images is not inherently more "superstitious" than many other practices such as selling indulgences or executing witches. Labeling unfamiliar religious practices as superstitious amounts to a refusal of trying to understand.

Since all religions involve assumptions about the reality of the supernatural, the boundary between superstition and reasonable belief is at best indistinct. In contrast to the present wave of antireligious rhetoric from scientists such as Richard Dawkins and Steven Weinstein, most of the educated and intelligent members of the human species from Plato, to Newton, to Einstein, have held some religious or metaphysical beliefs. This does not make religion "true" but it suggests that science does not require us to dismiss a large part of human culture. Finally, it needs to be pointed out that some seeming superstitions, such as the Chinese practice of setting off fireworks on New Years to scare off evil spirits are more cultural commemorations than superstitious beliefs. Omitting such observances can be discomfiting for some but this need not be fear of adverse supernatural consequences but simply feeling one has missed an important celebration of cultural membership. Sometimes the label of superstition is justified, for example in reference to harmful practices such as human sacrifice.

## RATIONALITY AND EMPIRICISM

Science and rationality overlap but are not equivalent because thought can be rational without being empirical. All science is rational but not all rational thought is science. In a certain sense, astrology is rational because it is deduced from a system of principles. The same is true of traditional medicine, at least those that use diagnostic techniques such as palpation of the pulse diagnosis or inspection of the tongue to diagnose illness based on the rational, though metaphysical, scheme of yin-yang and five phases, rather than on demonic possession or curses. Modes of thought that were once rational can cease to be so in the light of further knowledge; the geocentric universe of Ptolemy is a clear example.

Many practices that we now consider superstitious seemed entirely reasonable earlier in history. Some of these such as the *Yi Jing* (*I Ching*) retain a peculiar charisma even for rationalists. Here is the perspective of physicist-sinologist Ho Peng Yoke:

Fortune-telling is a subject that has received only mixed receptions in the past. In our modern age, some would brush this subject aside as either unscientific, or superstitious, but there do remain others who are still fascinated by it.
. . .
The *Ziping* (fate calculation) method does not use scientific instruments to observe temperature, air pressure, humidity and air movement as the weather forecaster does . . . but it is consistent with the same principles that explain traditional Chinese science. It does not involve the supernatural. . . . (Ho 2003:154f)

Divination, then, was rational within the limits of the understanding of the time. To believe in astrology in premodern China or India was to be in accord with the leading thinkers of the time quite different from believing in it today. It was certainly not due to ignorance.

From its probably prehistoric beginnings, all forms of divination from watching flight patterns of birds to reading tea leaves to interpreting the diagrams of the *Yi Jing*, assume that all events in the universe, even those that seem random, are part of a coherent pattern. This notion of coherence or interconnection, taken for granted in traditional worldviews, and modernized by the psychologist C. G. Jung as synchronicity, is for many a satisfying spiritual principle. It is also the root assumption of science that events can be understood in terms of underlying principles. However science diverges completely from metaphysics in the way it discovers causality.

In most of human history, prediction by means of stars and planets was orthodox, not countercultural as it is today. The *Yi Jing*, recently a staple of hippie culture, was endorsed by Chinese emperors. The motivations for believing in astrology today are complex but are not necessarily simple credulity, as its debunkers assume. Modern astrology utilizes the most accurate available data on planetary positions and has become dependant on computer calculations. This does not mean its prognostications are any more plausible than those obtained with cruder techniques. Its cultural status has, however, undergone a reversal because the attraction of astrology is in no small part due to a backlash against science, a resistance to being required to think in officially sanctioned ways. Most who use these methods probably do not take them entirely seriously. Like fiction, they involve a willing suspension of disbelief. More sophisticated forms of astrology as well as the *Yi Jing* retain a philosophical appeal that is separable from their employment in fortune-telling. Thus, Jack M. Balkin, Professor of Constitutional Law at Yale and author of a recent translation of the *Yi Jing*, says:

Although many people try to use the *Book of Changes* to tell the future, I believe this reflects a misunderstanding of the book's real value. The *Book of Changes* is best understood not as a fortune-telling device but as a book of wisdom that can help people think imaginatively and creatively about their lives. (Balkin 2002:5)

That Balkin and Ho, both eminent critical thinkers in their respective fields, find divination fascinating in itself, or as a source of self-understanding is quite interesting. Carl Jung was, as already noted, the first serious Western thinker to find value in divination, with his famous theory of synchronicity (Wilhelm 1967). Jung's theory is somewhat obscure but its basic point, that random selection of texts may evoke unexpected insights, is valid in the experience of many. In traditional China, reliance on the *Yi Jing* did not absolutely require supernatural beliefs, though many held them. The reason some held the book in such esteem may well have been the insights they obtained, rather than any reliance on any supposed prediction.

There is also a fantasy aspect to complex divinatory systems, which with their mysterious-seeming diagrams seem to promise answers to the most fundamental questions, if only we could properly understand or apply them. Diagrams too, have an aesthetic appeal for many. For this reason visual divination contrivances such as the *Yi Jing* hexagrams or the dials of the *luopan* can serve to stimulate imagination, just as do some religious art objects such as Tibetan mandalas. Divination seems to continue to meet human spiritual needs, and it is likely that many who use it do so in a skeptical spirit of curiosity. This does not mean that divination is in any way scientific, simply that it may have value for people who do not take it literally. Science does not require rejecting all aspects of culture that are not verifiable.

## WHAT IS SPIRITUALITY?

Defining religion or spirituality is far more difficult than defining science. Yet because "religion" was not a distinct category in the thought of traditional China or India, we must clarify what aspects of culture we can place under these labels. Two, non-Western religions lack some attributes that are assumed in the religions of the Book. "Religion" and "spirituality" do not have exactly the same reference. Usually, spirituality refers to personal qualities and experience, while religion refers to these in institutionalized form. Spirituality tends to refer to inner conviction and experience while religion refers to outward forms. These distinctions cannot be rigidly applied to traditional Asia. Much of what might seem religious to us, such as lighting incense as an offering to a deity, was done at home, at the level of the family. Rituals were carried out in temples and the populace flocked to them for religious festivals, but there was nothing comparable to the weekly church attendance required of Christians during most of European history. The term spirituality also tends to refer to the higher aspects of religion, such as morality, and to philosophical aspects to help make sense of the human predicament and, in some cases, to transcend it.

For purposes of considering religious and spiritual thought in China and India, I offer these defining characteristics, without making any claims of them being complete or having universal validity:

1) Doctrines defining how humans should conduct themselves in the world. This includes not only ethical rules but also social and ritual proprieties such as appropriate clothing for specific occasions and dietary teachings such as vegetarianism or avoidance of pork.

2) A cosmology describing the structure of the universe on both micro- and macro-levels, usually assumed to be congruent. This typically includes beliefs about supernatural beings, typically in a hierarchy with ghosts and demons at the lowest level and a God or gods at the pinnacle. Awareness of a deeper reality—Dao or Dharma—can be cultivated and this cultivation can transform human life, producing immortality as in Daoism or existence beyond all limits as in the nirvana of Buddhism. Cosmologies tend to include what might be termed an ethical geography, such as the ten hells of Chinese religion and the triple realms of Buddhism. A well-known Western version is Dante's *Commedia*. Included in cosmologies was the destiny of the person after death.

3) Practices intended to communicate with, or bring one closer to, the reality of gods or special states. These include ritual, meditation, mantra recitation and prayer, and exercises such as those of Daoism and yoga, which are carried out for physical or spiritual benefit. The goal of these practices may be a desire to be reach a higher spiritual level, or they may be for purely manipulative purposes as with curses and spells. An important motivation for ritual in Hindu and Buddhist Asia was to gain merit in the hope of a more favorable rebirth, or at least to lessen the chance of harsh punishment in this or a subsequent life.

Modernized versions of Asian religion intended for Westerners sometimes attempt to make them sound analogous to science by terming their practices, such as meditation and mantra, "spiritual technologies." This is misleading. Meditating on a mantra is not technology in the way typing on a computer is. Religious practices may utilize technology but are not technology in themselves. Thus when I listen to recorded chanting on a CD, technology is used but not what is spiritual.

4) The ultimate authority for determining truth resides in canonical texts and/or teachings of highly revered figures who are in varying ways both human and divine, such as Jesus, Shakyamuni Buddha, and Confucius. These teachings are assumed to be beyond doubt, though an extensive commentarial literature finds quite different meanings in the same words. The authority of these teachings derives from their supposed source; empirical verification is not involved, though some describe experiences such as visions, which constitute proof for them.

It is characteristic of traditional societies to look to the past rather than the future as the locus of both the greatest level of spiritual insight and the ideal society, though the past was often a construct to serve a later agenda. This is of course quite different from the idea of progress that is all but universally assumed in the modern world. In this way the premodern thought of India and China is contrary to the spirit of science. Science assumes that knowledge will be more complete in the future, though it does not necessarily assume that human life and society will be better.

5) Teachings that provide a sense of meaning and purpose to life. This may be a felt personal relationship to a deity, such as Guanyin, the Buddhist bodhisattva of compassion, or it may be more abstract, such as living in harmony with the Dao. Following such teachings gives a sense of living in a higher or better way. It may also contribute to worldly success. Life in accord with these teachings is held to be superior to all other ways of life and the way to attain one's highest potential as a human.

In considering the relation of spirituality to science in Asia, emphasis will be on cosmology or the structure of the universe and natural worlds, though at times we will consider other aspects. In the religions of the Book, spirituality is what happens in church and is concerned with God. Because Asian religion does not have required church rituals, the spiritual tends to be intermingled with the secular in people's daily lives. Religion was institutionalized but in a different way than in the West. Asian religion cannot be defined as that which is concerned with God or gods, though supernatural beings are often involved. In traditional Asia, religion, spirituality, and philosophy overlap. Thus yin and yang are part of the spiritual tradition of China, though they are not related to any sort of deity.

## RELIGION, PHILOSOPHY, AND ETHICS

With rare exceptions, philosophy in Asia, no matter how speculative, had the conduct of human life as its ultimate concern. It was assumed that a life conducted in accord with Dao or Dharma would be both ethically good and emotionally happy, though exceptions were obvious and required clever rationalizations. The correct way of life was also in accord with natural patterns. Though there were philosophical schools that advocated complete self-interest, these were a minor strain in Asian thought. Thus the legalist school held the sole concern of governments should be maintaining power, no matter how oppressive this was to the people. While this well described how emperors actually governed much of the time, the dominant Confucian school opposed it, at least in theory.

It is important to recognize how completely this vision of ethics differs from much of contemporary thought. The modern West is fascinated with transgression and self-interest. Philosophers of selfishness such as Machiavelli and Nietzsche have been highly influential, as are their popular avatars such as Ayn Rand. In the West, Indian and Chinese spirituality have become associated with the modern counterculture, but the hedonism of the New Age movement is not found in the Asian sources. While there have been transgressive thinkers in China such as Zhuangzi, who advocated avoidance of government service, they saw the solution to difficult times as withdrawal rather than protest or rebellion. Shakyamuni Buddha challenged many aspects of Hinduism including caste but his message was also one of withdrawal from society, not rebellion. The response of Chinese political thinkers to social problems was to extol exemplary behavior on the part of all; social change was not envisioned. Rebellion was usually inspired by charismatic leaders of heterodox religious cults who preached millenarian doctrines; notions of social reform were vestigial. When social change did come to Asia it was based on Western models, radical Marxism in China, social democracy in India.

The solution to social problems in China was assumed to lie in study of the ancient classic texts that formed the education of most of the elite and, of course, justified their privileged positions. It was never imagined that life could be improved upon, but it could be made worse if the ritual observances were neglected or social roles were violated. Such an attitude could not have encouraged discovery and innovation. Technological innovation did occur in such areas as agriculture and water management but the notion that such changes could fundamentally improve the human lot was absent.

## ETHICS, METAPHYSICS, AND THE STUDY OF NATURE

It is also important to note that the boundaries of ethically significant and ethically neutral behavior were drawn quite differently than in the West. It is only a slight exaggeration to state that every kind of behavior had ethical significance. Propriety required not only such obviously

ethically significant matters as observance of the Confucian five relationships, it also mandated using the proper sauce for each dish and wearing
clothes of the color appropriate for the season. (Of course we still have such
rules—ketchup is appropriate for french fries but not haute cuisine and
clothes are seasonal, though mainly for practical reasons.) Now failing to
follow such rules simply results in social embarrassment, not one images
that the natural order is transgressed. In China however, rules regarding
dress, food, and other minute details of daily life were felt to be necessary
for maintenance of the social and natural order. Excessive self-indulgence
by the emperor could cause floods and other natural disasters for example. Wrong positioning of a tomb would result in adverse consequences
to surviving relatives. In this sense, the natural and social realms were
responsive to each other. These beliefs were taken very seriously, silly
as they now seem. The economic cost of such rigid regulation was great
because much time was expended fulfilling ritual obligations instead of
focusing on practical matters.

The rules associated with Chinese cosmology formed a corpus of theory
and practice that has been termed by Thomas Aylward, " . . . the art of
scheduling and positioning," Aylward goes on to explain:

This art is predicated on the belief that the purpose of human existence is to act
harmoniously with the flow of time and position oneself symmetrically within the
stillness of space. A basic example of this is the knowledge that one should plant
in spring and harvest in autumn and orient one's residence southward to absorb
the warmth of the sun. According to the practitioners of this art, by studying the
movements of the stars and planets and the patterns of the earth's physical features,
we can discover when and where to act in all situations. (Aylward 2007:2)

A variety of divinatory techniques were employed to properly schedule
and position, all to a great degree based on the cosmology first recorded
in the early commentaries appended to the *Yi Jing*. Subsequently, these
divinatory methods became more and more elaborate, with detailed correction factors added. Wasteful as such efforts now seem, they indicate the
immense importance placed on ordering human life on what were taken
to be natural patterns. The intent here is not unlike that behind the use
of meteorology and geology to guide where to place buildings. However
the Chinese art of scheduling and positioning seems to have given little
attention to concrete observation, though it is possible that the practical
aspects were not recorded because they were thought less important than
ritual.

There were always skeptics, though their dissenting works have received far less study than those of the Confucian mainstream and they
never fully rejected the standard cosmology. For example, the Song

philosopher Wang Chong argued against the lavish funerals that were thought to be necessary both to demonstrate filial piety and to placate the spirits of the deceased. In doing so, he denied the existence of ghosts and an afterlife:

What sustains the life of a man are vitality [*jing*] and spirit [*qi*]. When a man dies, vitality and spirit disappear. When this happens the body begins to rot. Finally it becomes dust. Where could the ghost be? (Poo 1990:60)

This attitude is empirical; ghosts are improbable because there is no evidence to establish their existence. Chong opposed elaborate funerals as wasteful and the purpose of propitiating the dead as illusory. His was not the predominant attitude. As often the case with those who espouse what the majority does not want to hear, Chong's arguments were ignored. His comments show that some did see through the supernatural beliefs of the majority, though most skeptics probably kept their dissenting opinions to themselves.

## ETHICS AND SELF-CULTIVATION

In addition to the rigid social conformity of the Chinese art of "scheduling and positioning" there is a rather different theme in Asian ethical thinking: that self-cultivation is the means by which people improve their own behavior and thus benefit society. It is assumed in all Asian philosophies that self-cultivation is essential to a morally and spiritually satisfactory life and that overcoming illusion is the fundamental task for human spiritual development.

Even the extremely abstruse arguments of the Indian Buddhist philosopher Nagarjuna, which end up refuting just about every statement that can be made about existence, are not intended as mere display of logical virtuosity but to clear away the illusions that cause human suffering. Whether his arguments succeed, or whether they are of sufficient clarity to be understood by any besides specialist scholar-monks is arguable. However their intent is unquestionably to remove illusions so that people who properly cultivate can perceive things as they really are.

That Asian religion/philosophy promises to help us with our actual life problems is a major reason it has been so attractive to the modern West. Our own tradition has come to eschew such practical orientation. In contrast to Roman Stoicism, according to the British positivist philosopher A. J. Ayer, modern philosophy serves no function of offering solace or practices for self-cultivation. That classical philosophy in the West also emphasized personal cultivation is persuasively argued by a prominent French historian of philosophy (Hadot 2002). More often, contemporary

Western philosophers have dismissed Asian thought as less philosophical for just this reason: it offers relatively little pure speculation regarding the standard problems of Western philosophy such as permanence and change, free will and determinism, and so on. Another response has been to treat Asian philosophers as if they were really addressing the standard problems of Western philosophy, once their statements are transposed into more technically philosophical ones. This is the approach of Chad Hanson, among others (Hanson 1992).

## RELIGION, SCIENCE, AND THE CHINESE STATE

What we would now consider science was suppressed to varying degrees in China but not by religious institutions. Rather, imperial governments were always fearful that knowledge developing outside its control would be destabilizing. (This is not unique to China. Stalin, notoriously attempted to control genetics to conform to Marxist ideology and thus reinforce government power.) As will be covered in more detail in the following chapter, the legitimacy of the emperor, from which stemmed the authority of all administrative officials, was based on the cosmology of heaven, earth, and man. The emperor ruled by virtue of the so-called *tian ming*, mandate of heaven, evidence of which was the orderly functioning of nature: the succession of the seasons and, especially, celestial bodies. When natural patterns proceeded in their usual order, this was evidence that the emperor continued to enjoy heaven's favor. Any deviation from expected patterns of natural events could, on the other hand, be construed as heaven's adverse judgment of the emperor's behavior and so become a pretext for rebellion. Though there was certainly awareness of natural causes of disasters such as floods, the general assumption was that anomalies were due to impropriety on the part of the emperor. For this reason, independent development of scientific knowledge, especially astronomy, could cause loss of control of such knowledge by the imperial court.

The orthodox cosmology, though it could be turned against the government, was also essential to its legitimation. Since men of the educated class were nearly all officials, or aspiring to be, most had a vested interest in orthodoxy. When the standard cosmology was challenged, it could be construed as a veiled political message criticizing the emperor and be grounds for exile or execution. What now seem to be innocuous statements about heaven and earth or yin and yang could be interpreted as seditious. Thus anyone suspected of heterodoxy put himself at risk of terrible punishment. This did not completely prevent free exchange of ideas but inhibited it to varying degrees, depending on the current regime's intellectual tolerance. Nonetheless, one has the impression from Chinese literature and art that conversation among close friends was relatively free, though far more likely to be about literary and aesthetic matters than scientific ones.

Empiricism can be threatening to authority since its findings are not controllable. Examples are the British government's suppression of news about the bovine epidemic of mad cow disease and the similar initial suppression of information about SARS by the Chinese government. In both cases, attempting to cover up the problem led to much greater spread of the diseases. Similarly, global warming has worsened because of the cupidity of the petroleum industry and that of governments to ignore the evidence. Many politicians have difficulty recognizing the objective reality cannot be altered by rhetoric. This was no less true in the ancient world.

The suppression of knowledge outside the government's control took quite different forms than that of the Catholic Church in the West. There is similarity, however, in that both authorities were anxious about losing power and both jealous of their monopoly on the truth. Often it was not so much the specific ideas that were threatening but rather the notion that truth could be obtained independently of the respective authority.

Most revolutions in Chinese history, such as the Yellow Turbans of the Three Kingdom period and the Taiping Rebellion, which incorporated Christian elements, and even the so-called Boxer Rebellion against Western residents, were fomented by deviant religious groups. Typically part of the appeal of these movements was their espousal of an idiosyncratic revision of traditional cosmology. Thus Chinese governments have always watched enthusiastic popular religious movements with trepidation and continue to do so, as in the present persecution of Falung Gong. Because of the potential for religious leaders to stir up the populace against the government, emperors always tried to assert their authority over religious institutions. There was no clear doctrine of separation of secular and religious authority. Buddhist orders declared some degree of independence from imperial authority but in practice were obliged to submit to government regulation in varying degrees, which, for example, usually reserved the right to authorize ordination of monks. The growing influence of the Buddhist Sangha at times would lead to persecutions in which monks were forced to return to secular life.

## RELIGION AS THE OPPRESSOR OF SCIENCE: BEYOND THE MASTER NARRATIVE.

In the West, we tend to think of science as gradually triumphing over the beliefs of religion and even supplanting them. This view of the relation of science and religion is a "master narrative," that is, a story so persuasive that it tends to win uncritical acceptance. Master narratives are useful in that they help us put an immense quantity of facts—here Western cultural history—into a coherent pattern. Because actual history is not so neat, such narratives, if taken at face value, tend to inhibit nuanced understanding. The idea that the history of science is the triumph of truth over superstition began with the so-called Enlightenment of the eighteenth century. In its extreme form, as stated by Nobel Laureate physicist Steven Weinstein and geneticist Richard Dawkins, science is held to make religion obsolete. That most of the world's people still hold religious beliefs is explained away as regrettable ignorance.

The blossoming of spirituality in our own time shows that science does not replace religion. Rather, each is a distinctive mode of thought by which humanity understands its plight and plans action in the world. The most useful approach in their study is to consider the value each has for the people of a given time and place.

History, it is said, is written by the victors, and this has been particularly true with the history of science, which has emphasized what later turned out to be correct. Thus Copernicus and Galileo are heroes; Ptolemy, the codifier of ancient geocentric cosmology becomes merely an exemplar of error. More recent and sophisticated history of science goes beyond this master narrative and tries to consider earlier ways of thought in the context of their own time rather than as simply foreshadowing our own theories.

The master narrative has several weaknesses. Most obvious is that many prominent scientists from Newton to Einstein have believed in God. But a more important limitation of the master narrative is its inability to account for the rise, despite the accomplishments of science, of religious participation, as well as the revival of earlier modes of thought such as astrology, feng shui, Yi Jing divination, the concepts of yin and yang, yoga, meditation, rebirth, and traditional medicine. Far from being eradicated by science, these have become integrated into Western popular culture. Wittgenstein argued that prescientific conceptions and ritual should not be dismissed as errors but understood from within in terms of what they meant to those who believed and practiced them. This does not mean they have the kind of validity that science does but that they are more than ignorance. For many, they meet needs that science does not.

One reason to study the relation of spirituality and science in the traditional cultures of China and India is as a prerequisite to understanding their appeal for our own culture. However we must not imagine that current New Age conceptions of Asian spirituality are identical to those in Asia, either now or in the past. Hence another reason to study these phenomena is to be able to appreciate them in their own context.

## METAPHYSICS, MYTHOLOGY, AND SCIENCE

Because science is the dominant mode of understanding the world in our time, it is unavoidable that those who take religion seriously, whether as a subject of study, a set of beliefs or both, will feel a need to reflect on what the value of religious thought might be. For many, probably most believers, this need not be an issue. Personal religion provides a range of satisfactions such as cultural solidarity, a sense of purpose and place in the universe, reassurance in troubled times, relief of anxiety about death, social contact, rituals to commemorate important life passages such as birth, coming of age, marriage, and death. For those raised in a religion, it is simply part of their social and mental world and needs hardly more explanation than the daily rising and setting of the sun. An extremely influential view of religion is fideism, considering how belief meets human needs. While this approach is instructive—and has a great variety of approaches within it— it does not provide a complete account of the relation of religion to science. In addition to a fideistic or psychological or anthropological approach, it is necessary to evaluate the relation of religion and science in a more critical manner so as to consider the kinds of truth they discover and the ways they discover it. In this undertaking there are powerful, but ultimately limiting views that we transcend. To do this, we shall briefly consider the views of three twentieth-century philosophers:

The first is that of the logical positivists of the Vienna circle and their leading Anglophone exemplar, A. J. Ayer. Socially prominent in London circles, Ayer was a persuasive writer, though a rather dogmatic one. He found nothing of value in religion or metaphysics:

no statement which refers to a 'reality' transcending the limits of all possible sense experience can possibly have any literal significance; from which it follows that the labors of all who have striven to describe such a reality have all been devoted to the production of nonsense. (Ayer 1946:34)

This would sweep away, not only Western religious thought but also such Asian concepts as the Dao, Dharma, yin-yang, the three gunas, and many more. To a strict positivist such entities and any statements about them are unintelligible. The problem with this narrow view is that hundreds of millions of humans have, and still do, understand such statements.

The existentialist philosopher Karl Jaspers who was trained in medicine and became a psychiatrist, made the following remarkable statement:

More than once in history, metaphysics as a draft of being has assisted at the birth of scientific world orientation. The atomism of Democritus became a cognitive implement for modern natural scientists ... Kepler's astronomical discoveries originated in his metaphysical ideas of cosmic harmony. . . . In each case the scientific idea detached itself from its metaphysical roots.

. . .

Things which as objective metaphysics were . . . devalued into hypotheses, mere thought possibilities, only to rise with increasing verification, cleansed of all metaphysics, to the value of fruitful theories. Metaphysical impulses led to the creation of categories and means of thought . . . before they were secularized . . . and tried out as effective research tools. (Jaspers 1932, Vol. I:159)


Here Jaspers maintains the distinction between metaphysics and science but sees metaphysics in a positive light as a source of fertilization of science. His most striking example is the atomic theory of Democritus, in its time purely metaphysical but later essential to explaining the structure of matter. Where Ayer rejects metaphysics as unintelligible, Jaspers sees it as interactive with a scientific mode of understanding and often fertilizing it. Without denying the special truth claims of science we can avoid the unqualified rejection of metaphysics of the strict positivists.

This will be the approach taken in the present work. In considering science and religion in traditional China and India, a clear distinction between metaphysical and empirical thought will be maintained without the implication that metaphysics is per se inferior, or that it is a form of error to be outgrown. On the other hand, my approach here is not of complete relativism. If forced to take sides, I would admit that I do feel that science is a more powerful mode of thought than metaphysics. In my own field of biomedicine it has brought immense benefits including the ability to cure infection, relieve pain, and triple of the average human lifespan. Yet we can fully admit the power of science without rejecting all other forms of thought.

Significantly, Jaspers did not propose a rigidly evolutionary account but rather pointed to specific cases in which metaphysics stimulated scientific understanding. In traditional Asia, science consisted of episodic discoveries made by gifted individuals and groups rather than an evolutionary stream. Astronomy was something of an exception as it did show progressive addition of knowledge. Medicine, on the other hand, though it changed during Chinese history never seems to have improved its efficacy. Science in Asia did not fit the model of Thomas Kuhn who sees science as a sort of punctuated equilibrium in which conventional "normal science" is periodically shaken by revolutions that introduce a new paradigm in which the basic understanding changes (Kuhn 1970). Obvious examples are the Copernican solar system and twentieth-century physics with relativity and quantum mechanics. This model cannot be applied to the development of science in China and India because the paradigms in both civilizations were metaphysical and remained in place until the

Western presence. While important scientific discoveries were made in China and India, these were fitted into the existing cosmology and did not challenge the paradigm. To the extent that there were new paradigms, they were metaphysical and placed alongside the older ones instead of superseding them. It is dubious that scientific ideas in either India or China ever coalesced into a coherent paradigm; rather they were characterized by discreet discoveries and theories that tended to remain separate from each other. Nor did these necessarily inspire further research leading to more complete understanding. An evolutionary model applies far better to Asian technology, for example ceramics in China, which became progressively more delicate and refined than to theoretical science.

What did constitute paradigms in both cultures were the correlative cosmologies that accounted for all phenomena and also the proper ordering of society. These paradigms remained intact, though not without change, until the forced entry of Western powers. Then there was truly a new paradigm in which traditional scientific thought was replaced in toto by that of the West.

## IS RELIGION SIMPLY ERROR?

A brief statement of Ludwig Wittgenstein, arguably one of the most rigorous thinkers of the twentieth century, offers a corrective to simple dismissal of religious truth claims. In response to his reading of the celebrated study of "primitive" religion, Sir James Frazer's *Golden Bough*, Wittgenstein remarked, "Frazer's account of the magical and religious views of mankind is unsatisfactory: it makes these look like *errors*" (emphasis in original) [Wittgenstein 1993: 119]. In his subsequent comments, Wittgenstein displays little patience for the dismissal of religion as simply mistaken. While not maintaining that they are literally true, he points the way toward a middle ground of understanding. His most striking example is prayer to the Rain-King by a tribal society, a ritual that is performed only as the rainy season is about to start so that the participants know on some level that the ritual is not really what brings the rain (Ibid. 137).

Wittgenstein is telling us that we must look deeper than the surface implausibility of such beliefs. We must understand what they meant for those who believed in them. I have tried to adapt this perspective in this book. While it is foolish to regard the rain dance as of equal objective truth as modern meteorology, it is equally foolish to dismiss it as merely ignorant. Many believe, as Wittgenstein did, that studying the diversity of ways humanity has understood the world has intrinsic value and gives us a sense of connection to those who lived before us. Wittgenstein was not saying that the belief in rainmaking rituals is literally true but that within its cultural context it makes a certain sort of sense.

Within science, there are many now abandoned theories that were not simply errors. The geocentric universe of Ptolemy was a model that predicted many celestial phenomena with reasonable accuracy. The Copernican heliocentric system is more economical and capable of far greater accuracy. Yet to dismiss Ptolemy as simply mistaken would overlook the value of his model. Similarly, Newtonian physics is not mistaken but made more complete with the modifications of relativity. It is in this spirit that we must approach the science of traditional cultures, as modes of understanding that had considerable value in their own times, not as failed attempts at being modern science.

# Chapter 3

—

# Religion and Philosophy in Traditional China

## APPROACHING CHINESE THOUGHT

One of the rewards of studying other eras and cultures is becoming more fully aware of the range of human thought. Attaining this realization involves setting aside assumptions of which we may not be fully conscious, based on our own culture but which do not necessarily apply to all others. This is particularly the case when considering religion. Though many make the comfortable assumption that all religions teach the same thing, serious study of religion reveals as much diversity as similarity. For many scholars of religion, it is these differences that are of the greatest interest. In contrast, when considering science in other cultures, we look for manifestations that resemble science in the modern world; science everywhere involves the study of natural phenomena and testing of hypotheses by empirical means.

Though there have been claims that there are specific national styles of science; these are unconvincing, since knowledge can only be science if verified empirically. Even the social organization of science tends to be similar across geographical boundaries. Efforts to force science to follow ideologies, such as the disastrous agricultural experiments of the Great Leap Forward in China or Lysenko's genetic theories in the Soviet Union, have been conspicuous failures. In earlier times, however, scientific discovery was sporadic and lacked the institutionalization that now maintains continuity.

It is useful to begin by noting certain assumptions that are best set aside when considering the history of Asian science and spirituality:

- Though religion and philosophy have come to be distinct modes of thought in Western scholarship, this distinction is far less useful in considering Chinese and Indian thought. Confucianism, for example, has a resolutely secular focus, yet places great emphasis on ritual including devotional practices for the dead. Whether Confucianism is a philosophy or a religion has been debated since the Jesuit presence in China in the seventeenth century. This is a pointless debate because Confucianism has elements of both; philosophy and religion were not separate in traditional Asia as they have become in the modern West. While gods were not necessarily involved, self-cultivation was always central in Asian philosophy as it may well have been in Ancient Greece and Rome (Hadot 2002).

- Until recently, Western scholars of Daoism and Buddhism divided them into philosophical and popular or religious forms. The philosophical form of Daoism was that of the *Dao De Jing* and *Zhuangzi*. The magico-spiritual practices that were central to Daoism as a religion were assumed, incorrectly, to be followed only by the less educated. A similar distinction has been was made with respect to Buddhism in India and elsewhere. The intricate philosophical analysis association with that religion, and with Hinduism as well, was separated from devotional practices to religious images in the hope of receiving worldly favors.

These categories assume, dubiously, that those who studied the philosophy were not those who resorted to ritual or "superstitious" practices. Without doubt, many were interested only in religious practices operations only for the usual purposes of finding wealth, recovery from disease, protection against curses, and the like. Because these practices seem superstitious to us we assume that the educated saw them as we do. As has been pointed out, there were Chinese who doubted the efficacy of divination and the existence of spirits. But this doubt does not seem to have been widespread. Even Confucius did not deny spirits but simply counseled keeping them at a distance. We might guess that this was merely a polite way of expressing doubt about the supernatural but we have no evidence as to how Confucius actually thought about whether spirits existed. To assume that the more educated did not engage in religious practices that now seem superstitious is almost certainly incorrect.

Despite this, I do not suggest that we entirely abandon the distinction between popular and philosophical Daoism and Buddhism, simply that we recognize that the distinction we make was not made in the same way in traditional China. Indeed, maintaining this distinction is necessary in considering the relation of spirituality to science since it is the more philosophical aspects of spiritual and religious beliefs that can be related to science. Some, including at times Joseph Needham, have seen magic as a predecessor of science because it attempts to manipulate the natural world for human purposes and because it assumes the existence of unseen forces. This is misleading because it is difficult to find specific instances of magic progressing into science and because the unseen forces of magic and those of physics are entirely different. Electricity and radiation, demons and spells, though both may be deadly, are entirely different, not least because we can consistently measure the presence and activity of the former but not the latter. No doubt in the prescientific era, curiosity about the world could lead individuals to study both magic and science. However magical manipulations did not generally evolve into scientific ones.

Daoism offered systems of exercises involving breath, physical movements, and internal concentration that were held to be able to alter the internal structure of the body and confer longevity. These have, in simplified form, experienced a revival in contemporary Chinese culture and, indeed, have spread all over the world. There is a conspicuous lack of evidence that these exercises in fact succeed in lengthening lifespan. Contemporary masters of tai qi and qi gong do not have conspicuously longer life spans than the population at large. Chinese conceptions of health promoting physical activity did not include the notion of aerobic exercise. Thus, the fine motor activity employed in doing calligraphy, hardly aerobic, was held to promote longevity as were almost all activities the Chinese valued. Buddhist meditative practices could also be used for worldly benefit, even though they were probably first intended for spiritual benefit and accumulation of merit.

In traditional China and India, as now, spiritual activities and objects were part of the economy. The production and consecration of amulets and talismans, as well as chanting for the dead, were major sources of income for monks, though these entrepreneurial monks were presumably not the ones who produced the philosophical writings that hold the greatest interest for us today. The demand for such products and services was considerable. Most in Asia today associate Buddhism and Daoist establishments with magical and ritual services rather than with the philosophy that excites Western interest. Nor can we entirely explain away such practices as later degeneration of the traditions. In Buddhism, veneration of relics as well as production of amulets occurred shortly after the death of the historical Buddha, Shakyamuni, despite his explicitly forbidding such activities to monks. Since the earliest written records were compiled some centuries after the death of the historical Buddha, the practice of magic by monks may not have started as quickly as the texts would have us believe; nonetheless, these accretions began very early.

The best way to summarize this is to say that China and India had both abstract philosophy and practical magic and that the educated did not necessarily disbelieve in the latter.

• Neither Chinese nor Indian religion can be described as monotheistic. Indeed, both have a plethora of gods. It has been pointed out that in polytheistic religions, most individuals focused their spiritual activities on only one or a few deities. In India, there was near-monotheism in that most considered themselves followers of either Shiva or Vishnu; yet the existence of the rival god was not doubted. The notion that believing in one god requires renouncing belief in all others is mainly found in the religions of the Book. This does not mean, of course, that rivalry between religious organizations did not exist.

Buddhism in its early forms seems to have been unconcerned with gods because the teachings of the (human) Buddha placed him above the gods. Yet the existence of gods was not denied, only their importance. Given the human task of achieving enlightenment so as to escape the round of birth and death, it was the Buddha, not the gods whom one needed. In the Mahayana, the somewhat later form of Buddhism that predominated in China and the rest of East Asia, there are a near-infinite number of supernatural figures: Buddhas of past, present and future, Bodhisattvas, devas, apsaras, hungry ghosts and many others. This proliferation of divine beings is characteristic of Indian religious thought, though it readily took root in Chinese soil. A recent approach to religion, that of cognitive science, proposes that religion is about gods because the human mind inherently thinks in terms of beings rather than abstractions (Tremlin 2006:73ff). Many scholars of Asian religion would dispute the suggestion that it is mainly about gods. These religions do have elaborate philosophies that remain coherent when shorn of their purely supernatural elements. Tremlin is correct in stating that intuitive conceptions of gods as human- or animal-like are more easily processed by the human mind than abstract theological conceptions. However, spirituality is about more than gods. When we consider spirituality in relation to science, we are far less interested in deities and rituals than in how the religious beliefs depict the universe.

Because of the unrelenting monotheism of the religions of the Book, most of us have grown up with a cultural assumption that polytheism is inherently inferior to monotheism. A further taboo is the worship of "graven images" or idols. In both India and China, images of sacred beings are central to devotion and most worshippers do feel that the deity is in some way present in the image. (If this seems peculiar, we can consider the central ritual of communion in which the bread and wine are somehow really the body and blood of Christ.) The notion that polytheism is necessarily intellectually inferior to monotheism is simply missionary propaganda. While it is undeniable that science reached its full form in monotheistic cultures, if one traces the beginnings of the scientific attitude to such figures as Aristotle, Euclid, and Ptolemy, it is hard to argue simplistically that polytheism is less compatible with science, or with clear thinking generally, than monotheism. Indeed, no religion is scientific in that none ultimately base all their tenets on empirical verification. The existence of gods, whether one or many, can neither be proved nor refuted by science. Nor can we categorically state that monotheistic religions are less superstitious. Irrational beliefs cling to all religions, monotheistic as well as polytheistic.

- The gods of China and India were not conceived as all powerful like the God of Judeo-Christian/Islamic tradition. Nor was any deity assigned the three attributes of omnipotence, omniscience, and benevolence. Rather each god had a connection to a specific aspect of human life. Thus *Guanyin* (Sanskrit, *Avalokitesvara*) was the god, later goddess, of compassion and was called upon for overall protection or for aid when in danger. The most popular Chinese gods tended to be concerned with quite mundane aspirations: longevity, wealth, and the birth of (male) children. This serves to remind us that for most religion is concerned with all sorts of desires, even the most petty and spiteful, not simply those higher ones we now consider truly spiritual, such as world peace. There were also venal gods who had to be appeased with offerings. It is often suggested that Chinese envisioned the afterlife with a social structure like this world, particularly bribe-seeking officials. Indian gods were not always perfectly virtuous either. Krishna stole the clothes of cowherd maidens when they were swimming and Shiva tore of the head of his son in a fit of rage. (Such divine misbehavior is not unique to China—recall the sexual exploitiveness of the Greek gods.)

Buddhist deities were generally imagined as benevolent by both Indians and Chinese. Bodhisattvas, enlightened beings who chose not to enter nirvana but instead remained in samsara to aid suffering beings, embodied specific positive traits: compassion for Guanyin, as we have seen, cutting off delusions for *Wenshu* (*Manjusri*), healing for *Yaoshi* (*Bhaisajraguru*) and so on. They stood for spiritual qualities rather than gratification of daily needs, though people no doubt called upon their intervention just as they did the Chinese gods. Even the fierce gods of Buddhist and Hindu Tantra were ultimately concerned with overcoming desire rather than gratifying it.

Polytheism divides spiritual traits among multiple divinities. However, the spirituality of both India and China, did espouse universal principles that were regarded as sacred. The highest principle in Chinese spirituality is the Dao or Way; among its many meanings is the harmonious manner in which nature operates when its principles are followed. Just as humans can disobey the will of God in the religions of the Book, so they can be out of accord with the Dao, with harmful results. The notion of Dao assumes, as does the notion of God in religions of the Book, an inherent order that humans ought to respect and follow. Dao is usually translated as "the Way" both in sense of the way the universe works and the way in which humans should live. The Way is very much a natural entity, not a set of laws made by man. Daoism teaches that certain ways of behavior are conducive to well-being while others can bring one harm. Happiness is more a matter of living naturally than of obeying prohibitions. The Daoist conception of a natural life can seem passive in the sense of accepting things as they are. Confucianism also speaks of the Dao or Way but conceives of it quite differently from Daoism because it emphasizes is on social propriety rather than naturalness. In a way Daoism and Confucianism were complimentary, the former being concerned with the personal and the latter with the social. Most Chinese were influenced by both.

In India, the closest concept to Dao is Dharma, a somewhat harder principle to grasp. Within Hinduism, Dharma is something akin to law, rules of conduct for life. In Buddhism, Dharma is also often translated as "law" but really refers to the Buddha's teachings and the way of life that will enable release from suffering. To make an only approximate comparison, the Dharma of Hinduism is like the Dao of Confucianism, concerned with regulation of society, while that of Buddhism, like the Dao of Daoism, which is concerned with self-realization. In Vedanta, a philosophical development within Hinduism, the ultimate principle is atman, a concept that includes the Western idea of the soul but also the ultimate truth beyond individual limitations. For Buddhism the highest truth is sunyata or emptiness, a difficult concept meaning going beyond all conceptual limitations. In both Indian and Chinese religion, the most complete truth is beyond words.

The religious belief that there are underlying principles that are, at least in part, understandable by the proper mental effort or self-cultivation, can be likened to the tacit belief that underlies science as much as metaphysics: that the multitude of phenomena can be explained as manifestations of underlying principles. In the present era of fashionable nihilism, it is well to remember that all human activity assumes some relation to an underlying truth. Though the processes of discovery in science and in religion are quite different, both provide principles by which humans order their experience.

The closest China came to something like the Judeo-Christian God was *tian*, usually translated as "heaven," which refers both to the physical sky or cosmos and to an abstract ordering principle. Dao and tian, though impersonal and abstract, were more central to spirituality, at least for the literati, than any deity. Tian is the cosmic ordering principlethat the emperor in particular was required to follow. His right to rule was the *tian ming*, the "mandate of heaven;" loss of this mandate would result in dynastic overthrow. Another common phrase was (*tian yi*), the will, or intention of heaven. The highest form of wisdom for Chinese was to be able to perceive these intentions of heaven, something very similar to knowing how to accord with the Dao. Full understanding was what made one a sage. Good government meant that the Dao and tian yi were understood and followed.

- The boundaries between religions were far less distinct in China than in the West. The three major "religions" of China: Confucianism (which is only partly a religion), Buddhism, and Daoism, did compete with each other on a political and institutional level; but on a personal level, many, perhaps most, Chinese participated in more than one. This was not hypocrisy; it simply was not assumed that one religion must be chosen to the exclusion of all others. Because different religions met different human needs, many drew upon more than one. The gods of the Chinese, unlike Jehovah of the Old Testament, were not jealous—though they could be harsh and vindictive if not honored by proper ritual. However they seem not to have begrudged offerings to others, so long as they were not neglected themselves.

Politics of course affected the fortunes of religious intuitions. At times, for example, emperors influenced by Daoist or Confucian ministers, would dissolve the Buddhist monasteries and force monks to marry and return to lay life. Buddhism was particularly vulnerable to criticism for being of foreign origin.

In India, a great range of beliefs and practices were subsumed within what is now termed Hinduism. To some degree, the concept of "Hinduism" was imposed from the outside. In traditional India, people followed one or another set of religious practices but probably did not think of themselves as Hindu, at least until the arrival of Islam and other Western religions created self-consciousness about what religion one was.

## THE LITERATI OF CHINA AND BRAHMANS OF INDIA

It is conventional in sinological scholarship to refer to a particular social class as "literati." Though the term seems pretentious in modern usage, it has a quite specific meaning in the Chinese context: the literate class whose education prepared them for the civil service examinations that were the primary route of entry into government service. For this educated class, only government service held any prestige—merchants were despised, as were soldiers. Literati education was entirely based on the defined canon of Classics, although, as described below, this curriculum was revised from time to time and there were specific examinations for specialized knowledge, notably astronomy. Many literati developed interests beyond the classics; they formed the erudite physicians and scientists whose ideas and discoveries we will be discussing. Certain leisure pursuits were considered suitable for the literati and included poetry, chess, playing the qin (a seven-stringed zither), and calligraphy. Painting was also suitable. With these pursuits, a refined amateurism, even an affectation of clumsiness, was regarded more highly than virtuosity. Specialization occurred in the sciences and medicine but more on the basis of individual interest than any institutional structures. The literati tradition gave China a unique cultural unity but at the same time, the uniformity it imposed inhibited the development of specialized knowledge outside of the classics, including the natural sciences.

In India, there was also a literate elite whose studies focused on the classics, this was the Brahman caste, though members of other castes, particularly the Kshatriya, nominally the warrior caste but also that of rulers and officials, often received considerable education. Only Brahmans were qualified to carry out fire rituals and other essential religious ceremonies of Hinduism. Shakyamuni Buddha was a Kshatriya, indicating the heterodox status of Buddhism in India. In contrast to membership of the literati in China, which was, at least in principle, open to many, Brahman and Kshatriya status was hereditary. This social structure served to preserve knowledge of the sacred texts by mandating passage of priestly knowledge from one generation to the next, but restriction by birth must have had a somewhat stultifying effect on intellectual innovation. In both cultures, learning emphasized memorizing a limited number of ancient canonical texts.

## CONFUCIANISM, DAOISM, AND BUDDHISM: THE INTELLECTUAL AND SPIRITUAL TRADITIONS OF CHINA

Confucius is the name most indelibly associated with Chinese culture. The name is actually a Latinization of Kungzi, meaning Master Kung. Like Shakyamuni Buddha, Socrates, and Jesus, Confucius left no writings. The work associated with his name, the *Analects* (Chinese, *Lunyu*) best translated into English as *Discussions*, is a collection of the master's pronouncements on various occasions, usually in response to questions from his disciples. The date of composition of the *Lunyu* has received much scholarly attention; clearly some parts are earlier than others but the received text was put in its present form centuries after the Master's death, making it uncertain how much was really spoken by Confucius. Though the Master's pronouncements do give the impression of a powerful personality, most of what is ascribed to him is not present in the *Lunyu* at all but is derived from anecdotes in the *Mencius*, *Shi Ji* (*Records of the Grand Historian*), and other early texts. Some, perhaps most, of these stories are later inventions. Though he was credited with editing a number of Classics such as the *Shi Jing* (*Classic of Poetry*), and writing the "Ten Wings" of the *Yi Jing*, the dates of these works, as well as their style, show that Confucius could not have been their editor or author. Most of what was attributed to Confucius over the ensuing millennia has no textual basis but most does seem consistent with the spirit of the *Lunyu*. While no one doubts that Confucius is a real historical personage, as a factor in Chinese culture, Confucius is a composite, a convenient persona to represent a large cluster of interrelated ideas. The obvious analogy in Western culture is Jesus; most of Christian doctrine was never spoken by him, though Christians would be likely to hold that they are consistent with his message.

The very term Confucianism actually has no counterpart in Chinese. The equivalent Chinese term that it usually translates is *ru*, which designates members of the literati class educated in the classical canon. Those who would term themselves ruists would regard themselves as followers of the Confucian moral tradition, though not all ruists agreed exactly on its details. Ruism is a more accurate term than Confucianism in that it refers to a tradition whose founding was attributed to Confucius but which is based on multiple sources. Nonetheless, Confucianism is universally used in English and so I will employ it, rather than ruism.

Daoism, while its earliest texts were philosophical and explained a way of life that was both pure and pleasant, attained its popular appeal by its elaborate magic and its techniques to promote longevity, of which present day tai chi and qi gong are descendants. While the *Laozi* and *Zhuangzi* do not encourage desire, later Daoism caters to the human desires for sexual prowess and longevity. In its more philosophical forms it counseled withdrawal from social obligations, particularly government service, teachings that were anathema to Confucianism. The American sinologist John K. Fairbank influentially pointed out that the same person might be a Confucian and a Daoist at different times.

Buddhism came to China from India, by way of Central Asia. With it came a detailed cosmology, including hells, which seemed to quickly capture the Chinese imagnation and were adapted by popular Chinese religion. Buddhism offered several paths to relief from the sufferings of worldly existence, which ranged from release in this life using the meditative techniques of Chan (Zen) to rebirth in *Sukhavati*, a Pure Land presided over by Amida Buddha. While Zen and earlier forms of Buddhism required both extreme effort and virtuous behavior, rebirth in the Pure Land required, at least in principle, no more than reciting the name of Amida Buddha with the requisite sincerity.

Both Daoism and Buddhism were threatening to Confucians on both ideological and social grounds. In place of a life of government service, both provided ideological justification for escape from social obligations. Daoism recommended an eremitic existence in accord with nature, with an emphasis on preserving life rather than taking risks. Buddhism held out the promise of complete relief from suffering for those eschewing the life of a householder in favor of a life of intense spiritual effort. Both had cosmologies that reduced the social world to only one among several realities. Resolutely this-worldly, Confucianism condemned any way of life outside selfless service to the state. Buddhism was particularly despised by strict Confucians because of its foreign origin and its monasticism, which was contrary to the Confucian emphasis on reproduction as a filial obligation. We should not imagine that the followers of the three spiritual systems lived up to their ideals more than the followers of Western religions. Despite the self-sacrificing idea of Confucianism, official politics was ruthless with exile, castration, or death as the potential consequences of losing out in palace intrigues. Daoist priests often grew rich selling talismans and spells and Buddhist monasteries sheltered not a few "rice bag monks" who remained so long as they were fed.

Some, notably Joseph Needham, have proposed that Daoism encouraged science because it developed quite elaborate magical and alchemical procedures for manipulation of nature. Needham further suggested the Buddhist doctrine that everyday reality is illusory and unimportant except as a place from which to attain enlightenment, discouraged scientific observation. In the modern West, the opposite view has taken hold: Buddhism's near-infinite spaces and skepticism about the reality of the material world has been compared, misleadingly, to modern cosmology and quantum physics. In actuality, contributions to science were made by literati whose allegiances might be to any one, or more, of the three spiritual traditions. While many Western scholars consider Protestantism to have been a major factor favorable to the emergence of science in the West, neither in China nor India do we find persuasive evidence of a relation between religious affiliation and scientific proclivities.

While Daoism and Buddhism developed detailed models of the universe (cosmologies), these were creations of the imagination; observation played a very limited role in their formation. As products of correlative thinking, their main function was to explain the macrocosm by analogy to the microcosm of human experience. Seemingly practical techniques, such as the longevity-promoting meditations of Daoism, were based not on observation of what actually made people live longer—such would scarcely have been possible with the resources available—but solely on metaphysical theories developed from analogies. For example, since babies bodies are much more flexible than those of older adults, exercises to maintain flexibility, especially of the hands, are widely practiced by the elderly. However from a scientific viewpoint, stiffness is an effect of aging, not a cause. While such exercises may increase mobility and thus be beneficial to the elderly, as they are to the young, there is no evidence that they alter aging on a biochemical level.

## THE ANCIENT SAGES AND THE RISHIS

Many of the basic ideas of Chinese culture, including some practical inventions such as the plow and boat were attributed to legendary ancient sages or culture heroes, such as Fu Xi, usually shown bestride a leopard skin. Yao, Shun, and Huang Di were sage kings who ruled in a simple and honest way. Many later writers supported their own view of the proper ways of doing things by attributing them to these legendary rulers. The origin of important cultural elements, such as plows, are commonly explained by myth. However, the appeal of these stories may serve to satisfy the curiosity that might better have inspired a search for more factual explanations. Similarly, in India, the ultimate authorities were mythical ancient wise men called rishis. There was however a much greater tendency in India than China to attribute inventions to gods rather than legendary humans. In both cultures, the invention of culture heroes, whether divine or nominally human, may have made invention seem something beyond the abilities of ordinary men and women.

This retrograde approach to truth was, needless to say, not conducive to science. In science, what is of most interest and has the greatest credibility is what is most recent. New findings are questioned, of course, but it is assumed that science becomes progressively more accurate and complete. The validity of a hypothesis rests upon experimental data obtained with the most up-to-date techniques, not with its accord with ancient ideas. In science, though new discoveries may be ignored at first, more accurate explanations inevitably supplant the old. In traditional societies however, once the authority of the past is questioned, then the entire structure, both intellectual and political is threatened.

## SOME SEMINAL CHINESE TEXTS

Writing is at the heart of Chinese civilization. Unique among the world's writing systems, Chinese is logographic, that is the written language represents words in a nonphonetic manner. Recently, several leading scholars have insisted that Chinese represents sounds rather than words, ignoring the fact that Chinese speaking different dialects can understand the same written characters but cannot understand each others' pronunciation of them. This view of Chinese characters is an overreaction to the misconception that Chinese characters represent ideas directly without the mediation of language. However ancient characters can still be read with modern phonetics. Because the ancient forms of the characters can still be read three thousand years later, the writing system symbolizes the continuity of Chinese civilization. Additionally, in the form of brush calligraphy, Chinese writing is a form of artistic expression as well as a mode of communication.

### Oracle Bone Inscriptions

Most writings in China were on perishable materials such as bamboo strips. The earliest surviving texts were chiseled on to bone; the so-called oracle bones. These provide fascinating glimpses into Shang and early Zhou life, though not providing a complete picture (Keightley 1983). The discovery of the oracle bones is a romantic one. In 1899, the paleographer Wang Yirong, happened to see inscriptions on bones that were being dug up and sold as "dragon bones" for medicine. He recognized their similarity to early bronze inscriptions and began their collection and preservation. The oracle bones were employed for divination by applying heat to them to generate cracks. Expert diviners, and the king himself, interpreted these to help anticipate the events of the ensuing ten-day week. These texts include some references to astronomical events and assume a very simple cosmology in which divinatory queries and sacrifices are addressed to Shang Di, usually translated as "The Great God." Beyond this we have little idea of how they conceived the universe. Implicit in the act of divining is the notion of a hidden natural order that can be partially accessed by specific techniques, often with an element of randomness.

### *Yi Jing*

The *Yi Jing* or *Classic of Change*, more familiar by its earlier transliteration as *I Ching*, is the second oldest extant Chinese book, parts of the *Shi Jing* (*Classic of Songs*) being somewhat earlier. The *Yi Jing* is, together with the biblical books of Moses, the oldest book in continuous use. A universal resource for the scholar class from the Han onward, it has recently revived after suppression by the government of Mao Zedong, though Mao himself supposedly referred to it. It traveled to the West in the Richard Wilhelm/Cary Baynes translation, where it became an iconic text of the sixties counterculture. Carl Jung contributed a famous Foreword that introduced the notion of synchronicity and gave divination a rationale suitable for the twentieth century: consulting the oracle becomes a device for contacting one's subconscious mind. With this elegant translation, as well as many more recent ones, the *Yi Jing* continues to find new readers as one of the select number of spiritual works that somehow continually renew themselves with fresh modes of interpretation so that they seem never to go out of date.

In China, the *Yi Jing* was viewed both as a divinatory manual and as compendium of wisdom; it is of particular interest to us because it contains the earliest expressions of Chinese philosophy and cosmology. The *Yi Jing* is traditionally divided into two sections, the earlier, also referred to as the "Zhou Yi" is thought to have been composed during the early Western Zhou, somewhere between about 900 and 700 B.C.E. This early portion consists of sixty-four figures of six-lines each, termed *gua* in Chinese and hexagrams in English, each with brief accompanying texts. Each line of a hexagram may be solid or broken. Later the solid lines were associated with yang and the broken ones with yin. The texts consist of a title or tag, a brief text explanatory of the overall divinatory meaning of the hexagram and separate texts for each of the six lines. The origin of the hexagrams is unknown. The textual portion of the "Zhou Yi" is exceedingly obscure; this has given them an air of mystery and profundity and also permitted diverse and changing meanings to be read into them in accord with the personal outlook of the reader.

When we look at the *Yi Jing* text itself, we find mainly references to animals and other aspects of an early agricultural and militaristic society, including human sacrifice. A recurrent phrase, "favorable to cross the great river" reminds us of the very different human landscape of the Shang in which bridges were few and a boat trip across a river was a daunting undertaking. Another recurrent phrase, "favorable to see the great person" reminds us how dangerous was contact with those in authority. These earliest textual layers of the *Yi Jing* do not contain a developed cosmology.

The second section of the *Yi Jing* is a collection of appendices usually known as the "Ten Wings." Though traditionally attributed to Confucius, these were composed centuries after his death in the late Warring States or early Western Han, and do not resemble the *Lunyu* stylistically. Compared to the Zhou hexagram texts, "Ten Wings" are less opaque, though far from unambiguous, and more systematic than the Eastern Zhou portions. Most provide supplementary explanations of the hexagrams and their Zhou texts. In them we can see the beginnings of a philosophical sense quite absent in the "Zhou Yi." For the development of Chinese philosophy, the most important of the Wings are numbers five and six which form the *Da Zhuan*, or *Great Commentary*, the earliest textual record of what became the dominant Chinese cosmology. This cosmology was shared to some degree by Daoism and Confucianism, but with its achievement of official status in the Han, it is most closely associated with Confucianism. Also of interest in the *Da Zhuan* are myths regarding the origins of such early critical technology as the plow and boat (Wilhelm 1967:328–336). Not all elements of Chinese cosmology are present in the *Da Zhuan* but we can see most of the sprouts that later became trees. By the time of the *Huainanzi*, a text of the early Western Han, the cosmology begins to display the tendency for extreme elaboration, which continued until the modern era. While the *Yi Jing* is quite concise, in contrast Koh's manual of feng shui (2003) has nearly two hundred pages of elaborate numerical tables for determining favorable building arrangements.

Cosmology refers not just to the universe on a large scale but to its nature and structure at all levels: microscopic, human, celestial, and supernatural. Scientific astronomy and physics provide the cosmology of the modern world but the term often refers to conceptions that are religious in origin. Dante's *Commedia* is the best known example of a Western cosmological system but there have been others, for example that of the eighteenth-century visionary philosopher Emanuel Swedenborg. Prescientific cosmologies include not only the natural world apparent to the senses but also realms and beings we would now regard as imaginary. On the microcosmic level, there are metaphysical accounts of the basic composition of matter, such as the five phases in China and the three gunas in India. All cultures have cosmologies but they vary in complexity; not infrequently, as in China, multiple systems are conflated with the apparent inconsistencies overlooked. While cosmologies begin with simple observations, such as the movements of stars and planets, they tend to develop arbitrary elaborations. Though starting with what is visible, prescientific cosmologies seem to give the most attention to what cannot be seen, only imagined. A basic principle of traditional cosmologies is usually phrased, "as above, so below," a phrase often cited in Western esoteric writings. This principle holds that the structure of the microcosm, especially the human body and human society are parallel in structure to the universe on the largest scale. Thus the circumpolar constellations used in Chinese divination are also the emperor and his high officials. The principle of "as above, so below" is inherent in all forms of astrology which are based on the premise that the positions of astronomical bodies indicate what is happening, or will happen, in the human realm. Planets and constellations also determine situations and character traits. Thus Mars indicates a warlike temperament, Saturn governs the restrictions of life, and so on.

In actuality, traditional cosmologies develop in the opposite direction: "As below, so above." Human affairs are projected onto the heavenly bodies. Thus the appearance of a comet may be a judgment on the emperor's conduct. Cosmologies tend to be both descriptive—they delineate the supposed structure of both the human and cosmic realms—and prescriptive—they include rules to ensure that humans will live in accord with the natural order.

A Chinese equivalent of "as above, so below" was expressed by the Qing philosopher Lu Shih-I: "the principles of the cosmos are . . . the principles of the mind." (Henderson 1984:154f) This is the Confucian slant on cosmology in which the study of the universe is of value for the moral principles assumed to be inherent in it. Daoism also derived morality from nature but the lesson it found was being natural rather than following detailed rules. Later, Daoism carried "as above, so below" to an extreme in the form of extremely elaborate exercises designed to improve the internal structure of the body by meditatively placing the planets inside the body.

These two modes of correlation, one placing celestial structures within the body, the other seeking to understand the mind by studying the cosmos, share the assumption of macro- and microcosmic equivalence. Even now, some philosophers hold that because all we know is our consciousness, then the principles of the cosmos can only be the principles of the mind. But the Confucian view was different. Though perhaps a form of philosophical idealism but it did not question that the cosmos was "out there." It was closer to the Western movement known as natural theology in which study of nature was thought to reveal the ways of God. Natural theology in the West was, however, more empirical than the cosmology of China. In its effect on science, the Confucian reduction of cosmology to ethics could only be inhibitory. If the cosmos is considered of importance only for ethical self-cultivation, the chance to study it objectively as inherently interesting is lost.

Science, based on observation, has provided a cosmology that supplants traditional ones. The macrocosm consists of planets, comets, stars galaxies, black holes, and other enormous objects while on the microcosmic level, matter is made of atoms, which combine into molecules, which in turn account for the behavior of matter on the human level. At the same time, all of us intuitively express ourselves using the intuitive "as above, so below" cosmology. We assume that God is above us, we refer to high ideals and low morals, and so on. Science fiction offers physically impossible but easily imagined journeys into other worlds that resemble those of Daoism. Nonetheless, these are merely relics of traditional cosmology. To understand the way the universe seemed to those in traditional cultures, we must temporarily set aside our scientific model.

Traditional cosmologies are spiritual in the sense that they provide locations for religious entities and give human meaning to the celestial bodies, often associated with myths, while scientific cosmology depicts them as barren of life and meaning. Thus the Chinese imagined that the surface markings of the moon depicted a rabbit preparing an immortality powder. From the beginning, people have seen stories in the night sky.

The spirit of Chinese cosmology is already apparent in this famous excerpt from the *Da Zhuan*:

In high antiquity, when Fu Xi ruled the world, he looked up and observed the figures in heaven, looked down and saw the model forms under heaven. He noted the appearance of birds and beasts and how they were adapted to their habits, examined things in his own person near at hand, and things in general at a distance. Hence he devised the eight trigrams with power to communicate with spirits and classify the natures of the myriad beings. (Rutt 1996:421f)

This passage is followed by sections which describe the invention of some of the fundamental devices of civilization: markets, nets, boats, oars, plow handles, domestication of animals, gates, bows, ridgepoles, and coffins. Each is said to be inspired by one of the *Yi Jing* hexagrams. This much quoted and extremely influential passage is a good starting place to consider Chinese attitudes toward the cosmos in relation to science. The great culture hero, Fu Xi, is depicted as carefully observing the patterns of nature. These observations led to the formulation, inevitably reductive, of the general principles of the cosmos that are represented in the *Yi Jing* hexagrams. Indeed, the *Da Zhuan* states, of the hexagrams:

When we continue and go farther and add to the situations all their transitions, all possible situations on earth are encompassed. (Wilhelm 1967:313)

That is, the hexagrams are more than symbols, they are a complete account of the world. The claim by a spiritual book to provide all that anyone needs to know is not unusual. There is a natural human desire to find one source that explains everything. In the present case it expresses quite clearly the Chinese belief that truth is to be found by increased study of ancient texts, rather than in the study of phenomena themselves. This attitude is opposite to that of science, which certainly uses texts, though none have canonical status but is always seeking facts not already recorded. This reverence for their traditional writings, while important for cultural stability must have served to inhibit curiosity about the natural world in China and India. Yet this is not the whole story because other passages seem to credit observation of nature as inspiring discoveries.

In more than two thousand years of use, the *Yi* was used to support a great variety of ethical, philosophical, and political positions, often contradictory (Hon 2005). In the West also, ideas never conceived of in China have been read into this text whose beguiling obscurity renders it complaisant to almost any interpretation. One example is the similarity of the *Yi Jing* hexagrams to binary numbers. This was first noticed by the French Jesuit, Joachim Bouvet, who brought the *Yi Jing* diagrams to the attention of Gottfried Leibniz. Leibniz was intrigued by this similarity, which increased his admiration for China; but his awareness of the Chinese classic did not really contribute to the development of binary numbers.

The hexagrams do have some properties of arrays of binary numbers since each line has only two possible values, comparable to the 0 and 1 of binary notation. This limited similarity has inspired fantastical New Age works claiming to find computer science and the DNA double helix in this ancient text. Though the *Yi Jing* was the basis of innumerable numerological theories in China, its practical mathematical properties were never explored. As a binary array, the hexagrams are simply a curiosity; they do not add anything to modern number theory or to computer logic. The idea that the changes somehow anticipated modern mathematics or biology is purely fanciful and simply demonstrates how easily almost any idea can be read into this often maddeningly ambiguous classic.

The *Yi Jing* is also of interest as the text that most fully embodies the Chinese conception of time as qualitative, a characteristic of Chinese culture that will be further addressed in a subsequent chapter. As a divination manual, the *Yi* is less a collection of predictions than a guide to deciding the most favorable course of action at a given time. This notion that each time favors certain human actions has a powerful hold on the imagination—it is at the heart of Western astrology as well—and persists in contemporary China, where many bookstores have sections devoted to works on the *Yi*, more often referred to in China by its alternate name of "Zhou Yi," (*The Changes of the Zhou* [Dynasty]).

## The Other Confucian Classics

Traditionally there were six classics said to be edited, or at least approved by Confucius. One, on music, was lost early; the remaining five are the Classics of Change, Songs, Historical Documents, and Rites, and the *Spring and Autumn Annals*. The *Spring and Autumn Annals* are the historical records of a small state during the first phase of the Eastern Zhou Dynasty. It was commentaries that determined the meaning of these classics within Chinese culture. No classic had anything like a scientific outlook, though there are mentions of plants and wildlife in the "Classic of Songs," and some recordings of eclipses and other anomalous celestial events in the *Spring and Autumn* and its commentaries.

## CONFUCIANISM AND RESTRAINT OF THE SCIENTIFIC SPIRIT

In the *Lunyu*, Confucius displays little interest in the natural world. His concern is entirely the human realm and he sees society as a network of duties and obligations. These are summarized in the formula of the *wu lun*, or five human relationships, as listed by Mencius: love between father and son (the mother–daughter relationship is not mentioned), loyalty of subject to emperor, relationship of husband and wife, precedence of old over young, and faithfulness of friends (Yao 2003:664). Each of these relationships is hierarchical. Important in Confucianism is *zheng ming*, "rectification of names" expressed by the Master in a famous phrase that fathers should act like fathers and sons like sons. The exact meaning of rectification of names continues to be argued among scholars but here what is of interest is an implicit theory of language in which if words do not accurately reflect reality, then reality should be corrected. Put in a different way, words describe an ideal and the world should conform to this ideal. This conception has analogies in Plato's Republic and in the first line of the Gospel of John. It is contrary to modern philosophy, which places the problem on language, not reality. For science and in modern analytic philosophy, words and any sort of conscious representation are contingent on verification by observation of the external world.

The Confucian method of reforming society is to admonish proper behavior; we note that absence of anything like the modern ideas of social disadvantage or psychological abnormality as causes of harmful behavior. Nor is legitimacy of self-interest recognized; in this way Confucius's ethical philosophy resembles that of Kant. The reasons for improper behavior are not analyzed; the solution is moral admonition. The ineffectiveness of telling people what they should do is apparent. The approach of modern social sciences of trying to analyze the causes of deviance was lacking in China. The only alternative to Confucius was Laozi who held that if left alone by a virtuous ruler, people would be naturally moral. This is not much more realistic than Confucius's approach. The Confucian system of five relationships as a model of human society is an instance of a metaphysics onto which social reality is forced. The number five is preferred in China for numerological reasons (five phases, five notes of music, five internal organs, five colors, etc.) and so the complex variety of human interactions is reduced to five. This is not to say that more nuanced accounts of human interaction never arose but the Procrustean practice of reducing variety to only five categories inhibited development of more inclusive social models. With social roles, it was obligations, such as the wife's or son's obedience to father, and subject's to emperor that were emphasized. Many of what we would consider vital obligations, such as a mother nurturing her children were left out. In the *Lunyu*, Confucius seems to discourage speculative thought of any sort as distracting attention from proper social behavior. While he emphasizes the responsibilities of those in authority, proper leadership really refers to proper ritual performance. There is no notion of policy, for example of managing agriculture. The

Confucian official, at least as represented in these texts, is far from the modern technocrat.

The historical Confucius was born in what is present-day Shandong province. His life was spent wandering seeking an enlightened ruler who would engage him as an advisor but, except for a brief minor post, he was never offered a position. What might now seem like a failed career was interpreted as being due to his extreme rectitude, which would have been highly inconvenient to corrupt rulers. This is one of many examples in Chinese history in which the proper response to morally defective regimes was unwillingness to serve them. This was often making a virtue of necessity, since dissent could be met by extremely harsh penalties. Often an entire family was executed for the sole actions of one member.

At least among the educated, change in the system itself was never seriously conceived. There were rebellions, usually justified by the victor as in response to misbehavior of the previous emperor. The paradigmatic example in Chinese historiography was the overthrow of the last Shang emperor, who was famously self-indulgent and cruel. Revolution would overthrew one emperor and replace him by another, but fundamental change in the basic system of imperial government seems never to have been imagined until much later as a result of Western influence. Setting aside the mythic explanation that changes of dynasty were due to the mandate of heaven, revolt in China was motivated either by the power lust of warlords or the fanatical zeal of the founders of heretical cults. These later were utopian but never proposed practical reforms. Daoism did suggest governing by letting the populace alone, but this idea was never actually tried.

Any explanation for the persistence of China's despotic form of government over more than two millennia is speculative. Here it can be pointed out that correlative thought systems tend to support those in power, after all it is they who establish the systems. Because the emperor's role was taken to be ordained by the will of heaven, to criticize the governmental system would require questioning the entire metaphysical basis of Chinese culture. To "step outside the box" was virtually impossible.

## MENCIUS AND XUNZI

Mencius, who lived about two hundred years after the Master, is the most accessible philosopher of the Confucian tradition. While the *Lunyu* gives only hints of the Confucius's specific ethical principles, in Mencius we find a much more explicit moral philosophy. Separated from us by two and a half millennia, Mencius' way of thought is much more accessible to the modern sensibility. Even when disagreeing with him, his meaning is clear. Thus he notes how distressing it is to see animals being slaughtered for food, something which concerns many in the modern world. On the other hand his solution is far from the animal rights movement—it is to stay away from the kitchen, lest one's appetite be spoiled. Mencius is by no means insensitive but like other Chinese thinkers in the Confucian tradition, he could not imagine the conditions of life being much different than what they were. There was no doctrine of progress.

Xunzi, is also considered to be Confucian but famously differed from Mencius regarding the nature of human evil. Mencius considered mankind essentially good. Xunzi, in contrast, held that mankind is born evil and must be made good by education and ritual (Goldin 1999:7). As his disenchanted view of humanity suggests, Xunzi seems to us more a realist than Confucius or Mencius and, as we shall see, he had a skeptical bent.

## CHENG YI, ZHUXI, AND THE NEOCONFUCIAN SYNTHESIS

Over the ensuing dynasties, many commentaries were written introducing diverse interpretations of the cosmology of the Warring States and Han. Of particular importance were those of Zhuxi, Cheng Yi, and Shao Yong in the Song. Zhuxi is often considered the most influential philosopher in Chinese history besides Confucius, though his work was much influenced by his predecessors. Interested in Buddhism and Daoism in his youth, he had something comparable to a conversion experience as a result of which he declared Confucianism to be the proper way and maintained that the influence of Daoism and Buddhism must be opposed. His work is often referred to as the Neoconfucian synthesis because it integrated the moral teachings of Confucianism with some of the metaphysical ideas of Daoism and Buddhism.

Zhuxi's contribution was to assimilate some of the most appealing aspects of Daoism and Buddhism to Confucianism to improve its ability to compete in the cosmological marketplace. Though critical of the Shao Yong, the Chinese philosopher most noted for numerological interpretations of the *Yi Jing*, Zhuxi's extensive works included at least one treatise on numerology of the *Changes,* the *Introduction to the Study of the Classic of Change.* Neither philosopher's numerological obsession led to anything mathematically useful.

Zhuxi's version of Confucianism became state orthodoxy, a situation that was to last nearly a thousand years until the end of the Qing dynasty. Zhuxi was a profound and creative philosopher who integrated many disparate strands of Chinese thought. However as with any orthodoxy, Zhuxi's version of Confucianism became rigid and tended to confine intellectual activity to a narrow range, in great contrast to the creativity of the Warring States. Something similar happened during the European Middle Ages when Aristotle was the principal philosophical authority. In

Europe, the dominion of Aristotle ended in the Renaissance when alternative modes of philosophizing became available to the thoughtful and curious. Something similar happened in China with the arrival of Buddhism in the Latter Han, and again with the arrival of the Jesuits. Yet, while they created lasting changes in ways of thought, both were resisted as foreign and therefore inappropriate for the more civilized Chinese.

We should not blame Zhuxi for his stultifying influence in later centuries, any more than we blame Aristotle for medieval scholasticism in the West. Indeed, both systems, and Buddhist philosophy as well, continue to fascinate by showing what the human mind can accomplish by reasoning within fixed postulates. When a culture craves the security—and political stability—of fixed truth, it will find something to serve this need. An important effect of the Zhuxi orthodoxy was that it set the curriculum for the civil service examinations that determined entry to government service. The examination was based on the so-called Four Books: the *Lunyu*, *Mencius*, *The Doctrine of the Mean*, and *The Great Learning* (Legge 1892). Previously, the examination was based on the much earlier Five Classics. The differences between these curricula need not concern us here. What is significant is that the modified curriculum was still based on literary and historical texts. Though texts on astronomy, mathematics, and medicine became classics in their own right, they were never added to the canon for the official examinations. Many Chinese were, in fact, educated in these subjects but their attainment to high office was still based on the Four Books.

It should not be thought that this standardization of educational requirements was entirely deleterious. China is a geographically large country with considerable ethnic and cultural diversity. Despite this, it has maintained a striking degree of unity since original unification in 221 B.C.E. under the Qin first emperor. The agreement on the canon of ethical and historical writing was a major factor in this unity. Literate Chinese who lived a thousand miles apart had all read and memorized the same texts and shared a fundamental metaphysics.

# Chapter 4

---

# The Metaphysics and Cosmology of Traditional China

As we have seen, the term cosmology refers to mental models of the universe. All cultures have cosmologies because humans have always wanted to understand the nature and structure of the strange place in which we find ourselves. All cosmologies posit the existence of realms and beings other than those of ordinary waking experience. (This is equally true of scientific cosmology with is black holes and dark matter; realms probably stranger than the heavens and hells of traditional belief. Science offers no deities but holds open the possibility of extraterrestrial life forms.) Traditional cosmologies include descriptions of the destination of the soul after death. Shamanism, arguably the earliest form of religion, is based on the presumed ability of certain individuals to make virtual journeys to other realms, where vital information regarding problems of those in the normal realm can be obtained. The assumption that human affairs are partly determined by factors outside the human realm is thus quite ancient and only with the advent of science has it been seriously questioned. These other realms are discontinuous with the everyday one, but it is always assumed that travel to them before death is possible, at least for specially gifted individuals.

As cultures develop to the point where some individuals have the leisure to occupy themselves in philosophical speculation, conceptions of the universe become more abstract and complex. Thus there are multiple other realms, often invented to satisfy the human preference for symmetry; many cosmologies, notably the Buddhist one, posit equal numbers of realms above and below the human one. There is usually an ethical and affective hierarchy. The realms above this one in both Buddhism and Christianity are populated by happy, virtuous beings, while the evil suffers torments in hell below. China had its hells, in part derived from Buddhism, but belief in them seems to have been separate from its philosophical cosmology.

Characteristic of Chinese philosophy, generally, its cosmology was syncretistic, that is, seemingly unrelated systems are juxtaposed without a sense of them being not fully compatible. There are probably several factors behind this. First, the systems provided different sorts of explanations so that it was not expedient to abandon either. Second, although they may have originated separately, each retained authoritative status as no one felt entitled to decide between ideas attributed to ancient sages. This dilemma, of deciding between contradictory ideas of equally authoritative ancient teachers, is a recurrent one for societies that consider ultimate truth to reside in the words of ancients. No method is available to decide between authorities in apparent conflict. Often, there is a search for more accurate versions of texts in the hope that inconsistencies were due to textual corruption. This search, however, threatens those who derive their current authority from extant versions of the texts.

Science does not have this problem. The authority of Newton's laws depends entirely on their ongoing verification by experimental data, not on the prestige of Isaac Newton as a historical figure. Thus there was little difficulty in modifying them as subatomic physics made its dramatic discoveries in the first decades of the twentieth century. (There is a common misunderstanding that Newtonian physics was refuted by quantum physics and relativity. It was not. Rather it was shown that certain sorts of phenomena did not obey Newton's laws. That these laws accurately describe the sorts of phenomena that were observable with the science of Newton's time has never been in dispute.) Since traditional cosmology is metaphysical rather than empirical, there could be no commonly accepted means for modifying it. There could also be political resistance to change, given that the authority of the emperor rested on the mandate of heaven, that is, upon a cosmic principle. To question this was to threaten the basis of the entire structure of government and thus social stability itself. This also introduced a measure of vulnerability in that cosmic phenomena that were taken as confirmation that the emperor enjoyed the mandate of heaven might also be interpreted as signs of disfavor. The political implications of cosmic events were major determinants of the course of Chinese astronomy.

Contemporary cosmology is based on science, although much of the fascination felt by the general public for such conceptions as the big bang, black holes, the fourth dimension, quantum uncertainty, and "worm holes" connecting between parallel universes depends on their apparent metaphysical implications. While not lacking an empirical basis, such theories are not fully verifiable and stem as much from the human love for indulging in speculation about the nature of the invisible universe as from hard data. Such theories are now of intellectual interest only; since the eighteenth-century Enlightenment advances in understanding the structure of the universe have not threatened political authority. Religion has for the most part downplayed its outmoded cosmologies, no longer tenable in the face of the findings of modern astronomy. To the extent that cosmology remains, it tends to be reinterpreted in psychological terms, as in Carl Jung's explication of mythology as inherent in the structure of the unconscious mind. The modernization is not complete, of course. In the resurgence of "creationism" in American political debate, we see that many still feel a need for a cosmology that is religiously, rather than empirically, based.

## THE CHINESE COSMOLOGICAL SYSTEMS

There are three basic schemes in Chinese cosmology; over the course of Chinese intellectual history these became more and more intertwined. Simplest is the tripartite structure of heaven, earth, and man. This is inherent in the common use of the two character phrase "under heaven" to refer to the realm of ordinary life.

The three-level structure of heaven, humanity, and earth was simple and placed humanity in a clear conceptual location. Heaven was superior and set standards for proper behavior. In true macro- and microcosmic correlation, failure to live in accord with heaven's requirements resulted in disorder of the state and disease of the body. In the West, recent decades have seen a remarkable resurgence of the belief that living in accord with what is "natural" will maintain health and extend life. However, traditional Chinese notions of what is healthy or natural differed considerably from contemporary Western ones.

Translation of *tian* as heaven should not be taken to imply that this notion was at all similar to the Christian one. Heaven was not where one went after death. Rather it was an impersonal principle of natural order. Though associated with the sky and stars, and capable of satisfaction or dissatisfaction with the behavior of humans, tian was not anthropomorphic. In this, tian is quite different from the God of the Old Testament, who often felt and expressed anger. Heaven provided the Emperor with a mandate, which could be lost if he failed to observe a ritually correct mode of life. Going against the way of heaven somehow resulted in misfortune, but without anything like personal agency on heaven's part. We find something similar in the *Yi Jing*. Times are favorable or not, but this is independent of the emotions of the actors involved. On a personal level, however, personal emotion was often involved. Illness was often blamed on the curses of deceased ancestors who were bitter over not having been properly propitiated. This malevolence of the deceased toward the living was at least as often due to spite as moral disapproval. The dead or demons did not have a definite assigned location in the philosophically advanced cosmology of China. They were simply assumed to exist and to be able to adversely affect the lives of the living.

## DID THE CHINESE BELIEVE IN THEIR COSMOLOGY?

There is a tendency when writing intellectual history to imply that all members of a culture accepted all of its ideas uncritically. This was certainly not the case in traditional China. Here and there we find skeptical attitudes expressed in mainstream texts. It is likely that dissident views were underrepresented in texts in the classical Confucian tradition because these texts were selected for canonical status on the basis of their orthodoxy. As with the Judeo-Christian Bible, erotic and other transgressive elements were reinterpreted as metaphorical. Yet we find traces of skeptical opinion here and there. Thus in the *Zuozhuan*, an often cited commentary on early Chinese history, the following comment concerns some anomalous natural events:

These are the affairs of yin and yang; they are not what gives rise to fortune and misfortune. Fortune and misfortune come from men. (Goldin 1999:44)

In other words, it is humans who determine events, not metaphysical forces.

In a similar spirit, the Confucian philosopher Xunzi wrote:

If the sacrifice for rain is performed . . . and it rains, what of it? I say . . . even if there had been no sacrifice, it would have rained.

. . .

. . . we decide great matters only after divining with scapula and milfoil. This is not in order to obtain what we seek, but in order to embellish . . . Thus the noble man takes [these ceremonies] to be embellishment, but the populace takes them to be spiritual. (quoted in Goldin 1999:47)

This expresses a similar skepticism and suggests that divination and cosmological explanations of social and natural events were metaphorical rather than literal.

Throughout Chinese history, state control over interpretation of omens was necessary to control a generally credulous populace that did not doubt the reality of supernatural forces. That use of cosmological ideas for social manipulation was rarely explicitly acknowledged does not mean that officials were unaware of this. Thus Li Jing, a famous seventh-century general:

The art of war is an art of deception. Under the guise of yin and yang and *shushu* (metaphysics) the greedy and the simple-minded can be [easily deployed]. Hence they should not be abolished. (Ho 2003:10)

In other words, adroit interpretations of omens and portents can be used to build confidence in one's own soldiers and deceive the enemy as to one's plans. However General Li's comment does not necessarily imply that he believed that Chinese cosmology was wholly imaginary, simply that it could be readily employed for manipulation. We sense here the practical military man who has little use for abstractions unless they can help him to victory.

## THE DAO OR WAY

Dao (spelled Tao in the older Wade-Giles transliteration system) is the most fundamental concept in Chinese philosophy and is both metaphysical and ethical. The earliest use of the character simply meant a road or path. It soon developed a metaphorical meaning comparable to "way" in the sense of "way of life" in English. However Dao in Chinese is a more richly multivalent concept. It is the principle of the universe, the origin of the ten thousand things, the proper mode of life, and even the source of beauty. At its highest level, Dao is the ineffable principle of how the universe operates that is mainly accessible to the sage. This is how the Dao is presented in the *Dao De Jing* of Laozi and many subsequent writings down to the present day. To understand this Dao requires self-cultivation; direct or rational analysis is not sufficient. Effortful study can be self-defeating. Laozi tells us that in not looking out one's window one can know the whole world. It is implied that knowledge of Dao is the highest knowledge, far above mere knowledge of facts. This has sometimes been misconstrued as advocating ignorance over knowledge. Careful reading of the two early Daoist classics suggests that Laozi and Zhuangzi are not recommending ignorance but rather pointing to something beyond the limits of cognitive knowledge.

Dao has more specific meanings as well. Thus any skill or mode of knowledge can be termed Dao. Thus *chadao* is the way of tea; *shudao*, the way of calligraphy; *qindao*, the way of playing the zither; *daoshu*, methods of magic or sorcery; *daoxue*, philosophical study; *yang sheng zhi dao*, the way to nurture life; many other examples could be given. In each case, a skill is involved but also something more as the way of tea involves something more profound than simply brewing and serving tea; self-cultivation is involved. Similarly, calligraphy involves artistic expression that is far more than correct formation of the characters.

Although Daoism is often represented as opposed to Confucianism, the concept of Dao is just as central in the *Lunyu* and subsequent writings of so-called Confucianism. With Confucius however, Dao refers to morality and proper observance of social proprieties, matters scorned by Laozi and Zhuangzi. Both systems however share key assumptions that likely seemed too obvious to be explicitly stated: that there is an entity, not definable in words, but conventionally termed the Dao that comprises the principles of the natural work and human society. Knowledge of the Dao includes both an understanding of the natural or social world and of morality. In Chinese philosophy, metaphysics, and ethics, knowledge of the ultimate nature of things and of right and wrong are never regarded as separate.

Because philosophical Daoism places great emphasis on nature, some have tried to make a case that philosophical Daoism was a sort of proto-natural science or protoecology. This is misleading. In Daoism, the insight or knowledge of the sage was arrived at by an implicit process of self-cultivation. Empiricism was never central and nature was more a source of metaphor than an object of systematic study. However natural, the Dao was essentially metaphysical. Despite the search for a "theory of everything" in physics, science is not monist. It does not assume one underlying truth but rather discovers a multiplicity of truths. Hence it is doubtful that Daoism served as an inspiration for science in China.

As Daoism developed, it spawned elaborate practices concerned with increasing longevity. Magic was a prominent part of religious Daoism, particularly the production of fu, or talismans. Fu are modified Chinese characters with magical potency when shown, or even when burned and eaten as ash. To those with knowledge of normal Chinese characters, fu seem odd, even subtly ominous. The practice of fu assumes that the written word has a mystical power to affect events, particularly in the spirit world.

Many today find inspiration in Daoist philosophy; it has far more appeal outside China than the more somber moralizing of the Confucianism tradition and it adds a metaphysical dimension to existence that does not conflict with science.

## YIN AND YANG

The famous duality of yin and yang is hinted at in the *I Ching* but did not become fully elaborated until the Latter Han. Together with *qi* and *wu xing* (five phases), yin-yang was the most fundamental conception in Chinese cosmology or metaphysics. Yin-yang is a duality, perhaps the most famous one in all of the world's metaphysical systems; these two elements interact in a nearly infinite variety of ways to produce phenomena. Qi represents energy and, in some contexts, matter while the five phases are the component stages of all processes. These conceptions were extremely powerful in the sense of being able to serve as models for all phenomena. With science however, not all things are adequately described in terms of only two entities, nor do all processes consist of the same five stages. The problem is not so much that explanations in terms of yin-yang and wu xing were not scientific as that they satisfied curiosity too easily, reducing the motivation for more extensive investigation.

Yin and yang in China had quite different shades of meaning than the modern Western appropriation in which they are reduced to feminine and masculine. Originally, yin referred to the darker, northern side of a mountain; later the concept was expanded to encompass, among other things: darkness, cold, moist, hidden, invisible, female, interior, old, weak, unmanifest or potential, broken lines of the *Yi Jing*, white tigers, and ghosts. Yang refers to the opposites of these: warm, dry, visible, male, external, youthful, strong, manifest, unbroken lines of the *Yi*, green dragons, and the living. As with correlative systems generally, some phenomena seem to fit naturally into the yin-yang scheme when others do not. Heat and dryness do fit; hot surfaces tend to be dry because heat accelerates evaporation of water. Cold and moisture also tend to be found together because water condenses onto cold surfaces. Notice, however, that this scientific account of the association of warm-dry and cold-moist provides a causal mechanism but explaining that cold and moist go together because both are yin does not.

The sexual aspects of yin-yang are also experiential. The aroused penis feels warm because of increased penile blood flow while the female genitalia become moist, a change also because of increased blood flow. However attributing these to yang and yin tells us nothing about the physiological mechanism of arousal. Yin being potential rather than actual seems apt by analogy to pregnancy, though objectively women are no more or less potential than men. Other associations seem quite arbitrary, such as a white tiger symbolizing yin. As to ghosts and the realm of the dead, they are perhaps yin because hidden, yet this association seems to contradict the view of yin as potential. Yet the appeal of the system is apparent because many yin-yang oppositions such as light/dark, dry/wet, male/female, visible/hidden, young/old, etc., accord with how the mind naturally classifies.

Within science however, the categories of yin and yang have no meaning. There is no property of yin-ness shared by all the entities classified as yin which can account for wetness, coldness, darkness, potential, etc. Nor is there any objective test for classifying something as yin or yang. Though temperature can be easily measured, there is no evident criterion for deciding at which temperature yin becomes yang. In science, tempera-ture is a continuum while in yin-yang, metaphysics it is a qualitative dichotomy of cool and warm, based on subjective sensation. As to the yin world, the realm of the dead, it has so far been inaccessible by scientific method.

Yin-yang has become a popular concept in Western culture (though in modified form) because many find it pleasing as a metaphorical system. When we describe someone as very yin or yang, many would understand what is being conveyed. While this system is convenient for describing subjective attributes such as femininity or masculinity, it defines them only in circular fashion. Femaleness as a scientific category is defined by having two X-chromosomes, by presence of ovaries, and by female internal and external genitalia. As a matter of definition, we can label femaleness as yin but after this, further additions to the yin category are arbitrary. We cannot objectively establish that cool and moist have any property in common with femaleness.

Analogies are sometimes suggested between yin-yang and various con-cepts in modern science such as electricity. Needham suggested that yin and yang are analogous to positive and negative charge. This suggestion can be dismissed because nothing remotely resembling theoretical under-standing of electromagnetism is present in Chinese science. Furthermore, analogy to positive and negative misses a key aspect of yin and yang, that they are qualitatively different. This is not the case in electricity in which assignment of positive and negative was purely arbitrary. Posi-tive and negative in electricity have no quality analogous to the usage of these words in ordinary speech. The designation of positive and neg-ative with respect to subatomic particles could be inverted without any loss of meaning. Yin and yang however cannot have reversed attributes. The dead (yin), for example, are absolutely different from the living (yang).

Correlative systems like yin and yang lack the power of science because they do not add to our knowledge of the entities classified. This is readily apparent in medicine. To classify a symptom, swelling of the feet, for example, as being due a deep vein blood clot, or to heart or kidney failure points the way to further diagnosis and treatment in a way classifying it according to yin-yang and wu xing does not. Yin-yang and wu xing attributes are not really empirical ones. It is true that the dark and light sides of a mountain, to use the original meanings of yin and yang as examples, can be distinguished by observation as can whether an object is moist or dry, cool or warm. But there are no common qualities of yin and yang that can be directly observed in all objects assigned to each category, nor is there a nonarbitrary test as to whether something is yin or yang.

## QI

Initially *qi* referred to the vital energy of the breath. The character for qi depicts the vapor rising from rice as it is cooked, in analogy to the vapor visible with exhalation into chilly air. It has also been suggested that the character depicts mist over rice paddies (Cheng 2003a:615). The term also refers to human energy and vitality. It is quite close to the Indian concept of *prana* and the Greek concept of *pneuma*. Chinese medicine was particularly concerned with the circulation of qi, which took different forms and was held to travel through an elaborate system of invisible channels in the body. Like other metaphysical concepts fundamental to Chinese philosophy, qi was never disputed or abandoned though it did undergo considerable elaboration, particularly during the Song. Qi also had a moral dimension; Mencius held that a person's qi could be refined with self-cultivation so as to enhance ethical understanding and action.

Qi was held to exist in a variety of forms. Most fundamental was the division into yang qi, which was masculine, firm and upward moving, while yin qi was feminine, soft and pressing downward. This conception was used by Boyangfu to explain earthquakes as sudden release of yang qi that has been held down by yin qi (Cheng 2003a:616). An analogy to scientific geology can be superimposed on this since earthquakes are known to result from sudden slippage at a split between tectonic plates of the earth's surface. A quake then does result from release of blocked energy but to describe the friction that holds back movement as yin and the potential energy that builds up to the point of release as yang adds nothing to the scientific explanation. This is yet another instance in which an explanation that is partially scientific—that earthquakes result from release of pressure—went no farther because it was reduced to metaphysics.

Qi is a central concept in Chinese medicine. While it is tempting to consider it an early approximation to biological energy production during intermediary metabolism, the concept of qi provides no biochemical insight. While it is obvious that oxygen is essential for life, we cannot equate it with qi. Despite some suspect claims to the contrary, qi cannot be objectively measured. This is not to say that the concept of qi has no usefulness in the modern world. It is a convenient term, to refer to subjective energy. We may say someone does not have much qi when he seems listless. Many people distinguish male and female energy, and may refer to individuals as having lots of yin or yang qi. In such usage, the terms simply refer to something subjective that we feel about our self or others. Biological science has done little to clarify differences in human energy levels and loss of energy is a particularly challenging condition to treat successfully. The concept of qi helps us talk about such matters but does not contribute to scientific understanding.

In the Han, when the basic concepts of Chinese metaphysics were codified, qi becomes something more than simply energy. It was the life principle that becomes manifest through a process of materialization. This was made more systematic during the Song dynasty by the philosopher Zhang Zai who was versed in all three major traditions of Confucianism, Daoism, and Buddhism; his philosophy incorporated elements of the notion of void that is central in Daoism and Buddhism (Cheng 2003b:864–869). For him, qi existed in two fundamental forms: dispersed, in which there is no form, and concentrated in which qi exists as the objects of the world. This of course is simply an elaboration of Laozi's statement that all things arise from the Dao and return to it. *Li* or principle, another fundamental concept of Chinese metaphysics, is the way qi functions. The manifestations of qi are constantly changing due to the interactions of yin and yang.

Significantly for our inquiry, Zhang distinguishes between moral knowledge and knowledge "of seeing and hearing," that is, obtained by the senses (Cheng 2003b:867). However he prioritizes moral knowledge, which is obtained by cultivation of one's xin (best translated as heart-mind), over that obtained by observation. Thus the concept of qi, like all fundamental entities in Chinese philosophy, had ethical significance. The notion that observation of nature is mainly of value for ethical must have been another factor distracting from the objective study of nature.

## THE FIVE PHASES

The phrase *wu xing* that is now usually translated as "five phases" or, less often, "five agents" was once translated as "five elements." This latter translation is a misconception based on forcing the Chinese system into analogy with the Greek one. While the four elements of Greek metaphysics (earth, air, water, and fire) seem to have represented the constituents of matter, those of China definitely refer to processes, not matter. Traditional Chinese metaphysics was process oriented, not substance oriented, although qi began to be described in the Song Neoconfucian synthesis in terms that made it seem like matter in Western philosophy. However the matter-energy duality of physics never developed in China.

In the phrase *wu xing, wu* simply means five. The character *xing* has a variety of meanings, mostly related to "walk" but action is involved in all. Thus the five phases are quite different from the four elements of ancient Greek metaphysics because *xing* does not refer to constituents of matter. To interpret *wu xing* as "five elements" shows the danger of the once usual assumption that Asian concepts are just Western ones in exotic trappings. Sometimes earlier translators were aware that they were equating concepts that were not equivalent but assumed that Western readers would not understand more accurate renditions. If this was true once it is no longer.

The five phases are: wood, fire, earth, metal, and water. This is the so-called "mutual production sequence" in which each phase gives rise to the next in the series. There were other sequences representing different sorts of interaction between the phases, most prominently, a mutual conquest order in which each phase could overcome another. This was a basis for medical therapeutics. A water drug (cooling) might overcome a fire (febrile) disease, for example (Henderson 2003b:787). These sequences are explained by analogy. In the mutual production sequence, wood represents early growth; fire arises from wood, fire with wood produces ash, that is, earth; metal is produced from earth and can become liquid (as in bronze casting). In the mutual conquest sequence, the order was: wood, metal, fire, water, earth. In this cycle, each element comes into existence by the destruction of its predecessor. Wood is cut by metal, which in turn is melted by fire into water (liquid). Earth destroys water as when dammed by earth; earth in turn is susceptible to wood as with plowing (Lowe 1982:41). The whole system reminds one of the child's game of paper, stone, and scissors. Yet it developed into a system of considerable complexity that pervaded many areas of Chinese culture. The selection of five correlated with the five flavors, five musical tones, five internal organs, five directions, and so on. Organizing phenomena into numerical lists was equally common in Buddhist and other Indian thinking, though the significant numbers differed and the lists tended to have more direct religious significance. Buddhist lists, too, made omissions or introduced redundant items to make up the intended number.

It is easy to dismiss such simplistic systems as the five phases. Yet to do so is to expect that people of the ancient world somehow had the benefit of what science has learned since. Given the immense multiplicity of what must be processed by mind and senses, humans need to organize things systematically. Metaphor is a ubiquitous means of classification, even in science. Many medical terms are derived as analogies, for example the branching of nerve cells is referred to as "arborization." The difference, of course, is that the neuroscientist does not see any resemblance between neurons and trees except this very approximate one of shape.

### Examples of Correlative Classification into Groups of Five

Natural entities classified into fivefold categories included sensations such as colors, tastes, smells, and musical notes, emotions, internal organs, species of animals, meats, grains, and other vegetables. Man-made objects such as compasses and balances were classified by fivefold divisions. Social categories of five include states, government departments, administrative styles, locations for particular forms of sacrifice, even the room to be occupied by the emperor during each month. Celestial objects such as planets and asterisms were classified by five as were supernatural entities such as household gods.

Needham gives extensive lists of fivefold categorization (Needham SCC: II:262f, Table 12) as does Rochat de La Vallee (2003:157–189 and passim). Tastes were classified as sour (wood), bitter (fire), sweet (earth), acrid (metal), or salty (water). In theory at least, the best Chinese cooking arranges these tastes in a balanced manner. Smells were goatish (wood), burning (fire), fragrant (earth), rank (metal), and rotten (water). Animals were classified as scaly (wood), feathered (fire), naked (humans, earth), hairy (metal), and shell-covered (water). While one can see how some of these might arise by analogy—water can be salty, fish might feel rough like wood, shelled animals live in water, etc.,—there is no consistent principle by which entities are assigned to a particular phase. To be sure, animals are classified by their outer covering. But with tastes, for example, while we can readily recognize that water from the ocean and estuaries tastes salty, it is far less clear why earth would be sweet. With smells, the basis of the assignment is even less evident. Goatish odor is due to a particular substance, caproic acid, produced by the bodies of these animals, and is highly specific. Fragrances on the other hand are numerous and varied, as anyone knows from sampling perfumes or incense. They do share an affective quality of being pleasant but the other odors are not classified by whether they are pleasant or unpleasant. Goatish and rotten smells are both unpleasant. Many sorts of entities are simply left out. Most importantly, such categories do not facilitate finding of objective attributes as does science.

The principle of fivefold classification had important effects on Chinese culture. For example, it is well-known that China favored the pentatonic scale, though the septatonic was known and many pieces were composed in that scale. Because Confucius supposedly declared septatonic music to be impure as the tones deviate from propriety, pentatonic music was preferred. Here an aesthetic rule was derived from the preference for five.

## LI

Li was another central metaphysical concept in China but one that has not attracted the recognition in the modern world of yin-yang or qi. Li was explained in relation to qi, as an ordering principle, as qi itself gradually became something more like matter than energy. The material aspect of qi seems to have been a Song dynasty innovation as part of the Neoconfucian synthesis devised by Cheng Yi (1033–1107), his older but shorter-lived brother Chenghao (1032–1085) and Zhuxi (1130–1200). These philosophers were rather stern in their ethical views. As a youth, Cheng Yi announced himself to the emperor as an authentic sage. His self-righteousness only increased with age; his ethic for women was particularly unforgiving, for example, discouraging widows from remarrying. Zhuxi held similar ideas; ironically his own mother's success in managing the family household after the death of his father convinced him that women did not need to remarry.

Cheng emphasized li as the formative or ordering principle in contrast to qi as the active one (Hon: 2003:43–46). Cheng's school of philosophy was called *gewu* (investigation of things), a term that suggests learning about the external world. However investigation of things really referred to moral matters, not the functioning of the natural world. Zhuxi also gave li a central place in his metaphysics as an underlying principle. Li had multiple meanings including principle, reason, method, regulation, maintenance of order. In Zhuxi's usage it means something like the principle that determines things. What is ordered by li is qi. Things are what they are by virtue of Li. This approaches the distinction between form and matter in Western philosophy but it is important to recognize that such a distinction, while hinted at in some Chinese philosophical writings, never becomes explicit. There is never a conception of matter being acted upon by force, in the sense of modern physics. While it has been argued that the Chinese nonseparation of matter and energy is confirmed by modern physics in which matter and energy are related by Einstein's equation, $E = mc^2$, this is anachronistic and scientifically incorrect. The world's most famous equation does not state that matter and energy are the same, rather that they are interconvertible. Nor does there seem to have been any investigation of what qi might be in physical terms, beyond purely mental analysis. With li as the principle determining coalescence of qi into specific things, qi takes on a meaning closer to what we think of as matter, but without losing its energetic aspect. The concepts become murky as Zhuxi tries to explain them:

Whereas ch'i can congeal and operate, li does not have feeling, does not plan and does not operate.... *li* ... has no physical form or traces. It cannot operate. As for *ch'i*, it can brew, congeal, and aggregate to produce things. (Kim 2000:38)

Zhuxi also seems to use the concept of li to explain the problem of the one and the many. The Chinese version of this philosophical problem is how the Dao can be both separate from the myriad things and yet somehow be present in them. For Zhuxi, each thing has its own li but there is also a unitary heavenly li. By freeing the mind from desires, one can recognize the heavenly li and be in accord with morality (Kim:20). Li is thus both metaphysical and ethical. One can only speculate whether this nondistinction between ethics and the structure of the universe had any effect on the development of science. Plausibly, it gave a greater stake in the cosmology, which was seen as a support for personal morality and government authority. This is not unlike the concern on the part of some Christians past and present that questioning scriptural accounts of such matters as creation will undermine personal morality. There is no evidence that Zhuxi had any notion of distinguishing religious or metaphysical beliefs from empirical truths.

Li also was influenced by the Daoist and Buddhist ideas of the void or emptiness. These terms elude singular definitions. They refer to the hypothetical empty space from which all originates and to which all returns, and also to an ultimate realm beyond distinctions, both physical and ethical. That li as order or principle would be likened to the void may seem odd but in most Chinese and Indian philosophy, the ultimate principle of things was voidness. We find hints of this in the *Dao De Jing* in the notion Dao and of nonaction (*wu wei*). Emptiness is the central tenet of Mahayana Buddhism but is not quite identical to the void of Daoism, which is something like the potential from which everything emerges. In Buddhism, emptiness seems to have this meaning at times but more often is the ultimate truth beyond the world of the senses (Garfield 1995). Li as representing potential brought a central conception from Daoism and Buddhism into Neoconfucianism, enhancing the latter's philosophical and soteriological appeal.

However arbitrary it seems to us, its distinctive correlative cosmology gave traditional Chinese civilization a clear sense of belonging in the universe, something moderns might ponder with a bit of envy. Daoist and Buddhist philosophy have attracted a considerable following in the West in part because they provide, in somewhat different ways, a sense of the human role that does not depend on postmortem reward. In contrast, the rigidly hierarchical social views of Confucianism have made it of mainly historical interest. While Confucianism remains in the mundane world, Daoist and Buddhist cosmology seem to recognize both the limits of human knowledge and the presence of more possibilities in the universe than we can ever fully comprehend. Confucian metaphysics, given its primary concern with ordering society and justifying hierarchy, functioned as a closed system, implicitly rejecting any alternative system of metaphysics or ethics. Confucius's famous statement that he keeps his distance from spirits may be good sense but it is also a rejection of curiosity. The Confucianism paradigm tended to foreclose the possibility of other forms of knowledge and so cannot be seen as favorable to the development of science. The restriction was more mental than social—the educated were content with their culture's standard explanations and so did not attempt to go beyond them. To the extent there was institutional resistance to science, it was more secular than religious.

## DIVINATION

The earliest written records of China, the Shang dynasty oracle bones, were records of divination. Ever since, divination has had a central place in the life of traditional China. From emperor on down, all seemed to consult diviners for guidance in decisions, even ones that seem trivial to us, such as on what day to wash one's hair. In the modern world, all of us still seek expert guidance, though usually from professionals such as psychologists and financial planners rather than fortune-tellers. All of us are insecure in the face of important choices and not without reason, since many times in life we must decide among alternatives on the basis of incomplete information.

Divination assumes that things happen according to unseen forces that can be made apparent by special techniques, such as astrological calculations or selection of an *Yi Jing* hexagram, or directly perceived by individuals with special talents. All cultures seem to provide a role for such individuals. They may be shamans, who have access to special realms where hidden causes can be found, or they may be illiterate fortune-tellers, or highly educated savants with detailed theoretical knowledge. There is a spectrum from those having a mysterious special talent, such as shamans, and those who use techniques such as astrology or feng shui that anyone, at least in theory, might be able to acquire. Most often both talent and learning are required. The talent is partly supernatural but also seems to involve the mental function that we now term intuition. Intuition can seem almost magical; though its neurophysiological basis remains obscure, it is assumed to be an extremely rapid mechanism by which the mind processes large amounts of data so quickly that we are not consciously aware of the thought processes. The conclusion is presented to consciousness without the intermediate steps being apparent. Intuition is particularly important in human interactions—as when we have a sense of whether someone is honest or not—but less reliable with respect to the physical world—for example predicting the weather. To label this mental faculty "intuition" is not, of course, to explain it. Intuition is powerful but fallible, which is perhaps why technical procedures tended to develop to put divination on as rational a basis as possible. Diviners generally do not consider what they do to be intuitive but this faculty must be involved, certainly in telling powerful clients what they want to hear.

Smith, et al., point to a progressive rationalization of *Yi Jing* divination from presumed origins as a means of communicating with spirits to a way of connecting to impersonal qi and expanding one's mind in the direction of sagehood (Smith, et al. 1990:224). Zhuxi held that divination was the primary function of the *Yi Jing*, but in his view, this made it of more value than a purely philosophical or ethical text. While he was critical of Shao

Yong's numerological interpretations, he did draw upon this approach in his own works on the *Yi Jing*, particularly in his *Introduction of the Study of the Classic of Change* (Adler 2002).

It was Zhuxi, the great synthesizer of Chinese philosophy, who was most influential in integrating the divinatory and wisdom aspects of the *Classic of Changes*. By consulting the *Yi*, one would have a clearer picture of each situation, indeed, one would understand it in the same ways as the sages did. Such a notion is quite appealing—who would not want to be a sage and know exactly what to do in any situation? Nor has this appeal waned; many Westerners turn to the *Yi*, most often in the literarily excellent but not entirely accurate translation of Wilhelm and Baynes in search of guidance and wisdom(Wilhelm 1967).

Zhuxi attributed his philosophy to the mythical Fu Xi, presenting himself, not as an innovator but as the rediscoverer and preserver of the ways of the ancient sages:

...Fu-hsi, in a sense, prefigured Chu in the comprehensiveness of this vision. Fu-hsi created a numerological symbol system comprehending heaven-and-earth and human affairs, whose elements Chu Hsi integrated in his systematic treatment of moral-metaphysical, cultural, and numerological questions. Chu seems to have seen in Fu-hsi an incipient reflection of his own synthesis. (Smith, et al. 1990:224)

This tendency to see one's ideas as a revival of the ways of a happier past was nearly universal among Chinese philosophers, though those concerned with empirical knowledge often remarked that their sort of knowledge had expanded since ancient times.

Divination assumes a rational universe that operates according to regularities, if not by explicit laws. If events were truly random, they could not be predicted or their causes determined. In the prerational era, the causes were typically supernatural and emotional—intense envy as the motive for a curse being placed. It was assumed that negative feelings could have actual physical effects. Divination was based on somehow detecting the course and counteracting it. However divination can be based on rational principles in the sense that it is reasoned out from established rules. We tend to equate science and rationality; however they are not identical. Science must be rational but rational arguments need not be scientific. For example, medieval arguments for the existence of God, such as the unmoved mover or the ontological argument, are rational but not scientific since they are not based on empirical observation. Divination, however, can only impede the progress of scientific understanding of nature, not simply because it misattributes causality but because it reduces nature to a reflection of human concerns.

The Chinese considered the multiplicity of the world as arising from the interaction of specific forces that could be described and understood.

Most familiar in the West is yin and yang which have entered popular metaphysics, in no small part because they fit conveniently with the modern obsession with gender. However, feminine and masculine are only parts of the yin-yang system. It is assumed that all situations are due to particular interactions of yin and yang. To reduce everything to two underlying factors lacks explanatory power and, inevitably, the system was elaborated, especially in medicine in which each of the major organs had its own form of yin and yang.

While social position in India was largely hereditary, in China it was acquired, though the wealth and status of one's family greatly affected opportunity for high rank. In both cultures, ritual obligations accompanied social position and were of great importance. Whether rank was determined by birth or by examination and appointment, it largely determined permitted and required behavior. In China, since eligibility for official appointments was established by examination, abetted by political maneuvering, an individual could rise (or, often, fall) in social rank. Changes in rank then entailed changes in ritual obligation. As John B. Henderson (1984:57) points out, for the Chinese the cosmos was not essentially hierarchical, but human relationships were. One's expected behavior as well as privileges and obligations depended upon the formal social positions. A man would be subservient to higher officials but authoritarian to his sons. Women were admonished to obey first their fathers and then their husbands. This system did not allow much for flexibility within categories. Though it was sometimes suggested that the son's filial obligations might be waived in the event of improper behavior by the father, the emphasis was always on the rules, not on exceptions to them. In effect, the scope of allowed behavior was decided primarily by the category to which one belonged, rather than to individual factors. As always, of course, the ingenious found ways to escape social restrictions.

Though it would be simplistic in the extreme to blame the yin-yang metaphysical system for the poor treatment of women in traditional Chinese society, it may be presumed that inclusion of femininity in the same category as weakness, scheming, ill health, and death did not help matters. Correlations were also embodied in social policies in other ways as well. Thus spring was considered a time of happiness and so punishments were more lenient than in the fall. Yet by any reasonable ethical theory, punishment should depend on the nature of the crime, not on the season in which the trial takes place.

## NEEDHAM'S PROBLEM

Joseph Needham (1990–1995) spent his academic life at Cambridge University. Trained as a biochemist, he later switched to the study of Chinese science. His *Science and Civilization in China*, nominally seven "volumes" but actually seventeen thick tomes with more on the way, is an achievement unparalleled in any field of scholarship. Much of the work was done by collaborators, yet it was Needham's leadership and inspirational talents that made it possible to recruit the leading scholars of Chinese science for this Promethean task. Not only was Needham fluent in Chinese and widely learned in its culture, he had a lively and witty style. Though often opinionated, Needham always had a basis for his views, though inevitably some have been superseded by subsequent research. Because of *Science and Civilization in China,* we have a far more complete knowledge of science in China than we do for any other traditional culture.

Needham had an admirable enthusiasm for the achievements of Chinese culture. Yet at times one finds a defensive tone, a preoccupation to make his beloved China seem as scientific as possible. Some of his claims, such as the supposed discovery of the circulation of the blood, were based on what can be described tactfully as selective readings. Needham was preoccupied with what is now referred to as "Needham's problem" which is why science in traditional China, given its immense creative achievements, did not blossom as did science in the West. Contemporary modern Asian scientists have produced discoveries equal in importance to their counterparts in the West, but they were trained in Western science after its adaptation in Asia; their work did not derive from their native traditions.

Questions like Needham's are out of favor now because to ask why another culture is not like our own seems to imply that it should have been. Furthermore, as is often pointed out, it is problematic to try to determine why something did *not* happen. Yet, Needham's problem seems impossible to avoid as we look not only at Chinese science but also at that of India and other traditional cultures. To fully appreciate the science of the traditional world, we must consider how differently people thought. This does not mean a complete relativism that ignores the far greater explanatory and predictive power or modern science. It is fashionable, for example, to hold open the possibility that traditional Chinese medicine may be as effective as the modern scientific form. All that is required to disabuse us of this naïve view is to compare the present life expectancy in the developed world to that of Qing China, which was less than thirty. It is plausible that traditional Chinese medicine, or Indian Ayurveda, will be a source of effective remedies so far undiscovered in the West. Yet as systems, they are undeniably less effective than the modern one.

Recurrently in China and India brilliant discoveries were made but their potential was not realized. Standard examples are gunpowder, the compass, and smallpox vaccination. In general, science in China was not cumulative. Modern science, contrary to the romantic view in which the important work is done by a few geniuses, consists of the accumulation of small bits of knowledge discovered by the immense labors of thousands of scientists. The most brilliant discoveries are not born suddenly in the scientist's head but begin with consideration of the work done on the same problem by others. Results of research are rapidly shared by publication in international journals; nothing comparable existed in the traditional world.

There are important exceptions, however, such as the development of ceramic technology, which, as described in a later chapter, far exceeded anything done in the West both technically and aesthetically. Indeed, the appearance of the elegant Western porcelains of the eighteenth and nineteenth centuries was possible only because of import of Chinese technology. The technical advances, apparent to anyone who views an exhibit of Chinese porcelain arranged chronologically, are likely due to the fact that manufacture was carried out at a few kilns so that innovations could be passed on to others working in the same location. In tightly knit communities, oral transmission of discoveries could be effective, but it could not have been across all of China or India.

At least four theories have been advanced to explicate Needham's problem. First, the Chinese correlative system, by providing orthodox explanations for all observations, tended to inhibit the search for more empirically based ones. The notion of the will of heaven, by suggesting that nature somehow reacted to human ethical impropriety similarly inhibited the development of mechanistic explanations. Needham himself placed some of the blame upon the influence of Buddhism with its teaching that the world revealed by the senses is illusory. He suggested that by devaluing observation of the natural world, this tended to inhibit direct study of material reality. By disdaining the everyday world, Buddhism supposedly distracted interest from it. The problem with this view is that disdain for the world was equally prominent in Christianity, at least through the Renaissance. One of Jonathan Swift's satires was entitled *A Panegyric Upon the World* as if the very idea of praising the world was itself absurd. Science emerged in the West *despite* the disdain of Christianity for the world. The notion that Buddhism inhibited science is a sort of intellectual history that has become obsolete. We no longer imagine that ideas directly cause events. It is true that they have much influence on human behavior but to be plausible, there must be evidence that the idea actually caused the event (or in this case, nonevent) in question.

A related theory holds that the language of traditional China was not suitable for expression of scientific precision (SCC VII:2:95–199). Classical Chinese is essentially that of the Warring States period when the canonical texts were composed; though it underwent subsequent change, it did so far less than did the vernacular language. Classical or literary Chinese, like Latin in the West, was the language of the learned and remained the literary language until the twentieth century. Before that no writing would be taken seriously that was not in classical form. Classical Chinese not only lacked indication of tense, number, and case but was also a language of extreme concision, and therefore highly ambiguous. When classical Chinese texts are translated into modern Chinese, about twice as many characters are typically required. This is because clarifying words must be inserted and because words in the modern language are usually represented by a compound of two characters, allowing greater specificity of meaning. Since written Chinese is without inflections, the grammatical role of a specific word is entirely dependant on its position in the sentence, though there are characters that serve as grammatical markers. It is sometimes unclear whether a particular character is meant as a noun or a verb, for example. This ambiguity may be a positive attribute for literary use, but not for science. The question of how the Chinese language affected critical thought is an important one and is covered in more detail below.

A second theory for the incomplete development of Chinese science is that of Toby Huff who has claimed that East Asia and the Islamic Middle East lacked the institutional structure necessary for the development of science (Huff 1993). In Europe this structure was the university, which continued on in perpetuity and so kept ideas and discoveries alive in both classroom and library. Though this theory has been much attacked, it seems to me basically correct. Science must be cumulative, that is, it must have a mechanism by which new knowledge is added to existing knowledge. This requires institutions that teach the young and also provide a location for exchange of data and ideas. While China did have academies, these tended to be relatively short-lived. Certainly there was nothing like the Sorbonne, Oxford, Cambridge, Heidelberg, and other European universities that have operated essentially continuously for more than a millennium.

A third further factor inhibiting the development of science in China was imperial suspicion of intellectual activity independent of court control. Autocratic governments always feel under threat because their legitimacy does not rest on any meaningful consent from those governed. This factor is related to that proposed by Huff because imperial suspicion of learned bodies inhibited their establishment and prevented their perpetuation.

A fourth possible cause for the limitations of science in traditional China is the orthodox cosmology. In this work I emphasize cosmology as the inhibiting factor because it represented a fixed system of explaining phenomena that would have been threatened if empirical discovery had been pursued to its fullest extent. The traditional cosmology persisted for at least two reasons: First, it was more effective than it should have been in satisfying intellectual curiosity and second, challenging it would threaten the entire social and political system.

These hypotheses explaining the incomplete development of science in China are not mutually exclusive. All probably played a role. In this account, we do not isolate opposing factors as Needham did in suggesting that Buddhism inhibited Chinese science while Daoism stimulated it. There was no manifest rivalry between science and religion. Rather, Chinese intellectual leaders saw no necessity for thoroughgoing empiricism.

In contrast, in Europe a major stimulus to the development of science was natural theology, the doctrine that study of nature would reveal God's laws and thus confirm the truths of Christianity (Barbour 1997:17–29 and passim). While the relation of natural philosophy to Christian doctrine was contested, many did hold that it was a valid way for humanity to further understand God's ways (Olson 2004:84–103 and passim). For many Christians, God's revelation is something that unfolds in history. Chinese spirituality however tended to assume that the necessary truths had been determined and that study of authentic texts or with a truly enlightened teacher would reveal them. It was taken for granted that nature was adequately comprehended within Chinese cosmology; investigation of the external world had value but tended to be taken as part of the traditional metaphysics rather than modifying it. Fundamental truths were to be discovered by looking inward, not by empirical observation. Thus that famous line of the *Dao De Jing* that a person could know the entire world without looking out the window. If we also recall that the unchanging Dao is not the one that can be described, we get a sense of the Chinese approach in which the highest knowledge is instinctive or intuitive, rather than a series of factual propositions. This is no less the case with Confucius, for whom knowledge of the rites (proper behavior) is also intuitive. Indeed for Confucius, the highest are those who are born with knowledge; those who have to learn it are on a lower level. Nonetheless, the Confucian tradition has always placed extreme emphasis on education. The more skeptical Xunzi held that humans are born evil and need to learn to be good, but this view did not predominate. With science, of course, knowledge is not present at birth but must be learned.

## LOGIC AND THE CHINESE LANGUAGE

It has also been suggested that formal logic was never fully developed in traditional China. While Needham argues that all forms of logical argument can be found in Chinese writing, and that the structure of the Chinese language is more rigorously logical than that of European languages, as he puts it, " . . . the Chinese were always more interested in the truth on which assumptions were based than on the verbal machinery for developing these assumptions. Explicit logic did not therefore have that continuously sustained interest which it has received in the West." (SCC VII: 1:xviii) Though the Chinese were greatly concerned with writing and language, meticulous analysis of the sort that has been prominent in Western philosophy was never a focus. As Needham and others have noted, formal logic did develop in China in the Mohist school; this philosophical school did not persist after the Warring States, in part because it was associated with the artisan class rather than the literati class whose ideology came to dominate the official culture. The crucial point made by Needham and others is that when translation of Indian Buddhist texts into Chinese required development of a system of logic, the Chinese language was able to accommodate this. Formal logic was highly developed in India, though in a quite different form than that of the West, as it mainly served religious reasoning (Ganeri 2001). Logic was central in Mahayana Buddhist doctrine, particularly the work of Nagarjuna, the philosopher of emptiness (Garfield 1995).

Christoph Harbsmeier (SCC VII:1:399) demonstrates how early Chinese translators of Indian works on Buddhist logic were able to make precise Chinese versions simply by inventing their own technical terminology. Chinese does not have different noun forms for singular and plural while Sanskrit does. Translators resolved this problem easily by using a special character before certain nouns to indicate plural number. While this violated traditional classical Chinese style, it did produce the needed precision in meaning.

As this small example indicates, Chinese could modify their language when they wanted to express new ideas, though, as in the case of Buddhist sutra translation, the impetus was often the wish to render foreign ideas into Chinese. Nor should too much be made of the ambiguity of classical Chinese because the same is true of the canonical texts of many cultures, including the Judeo-Christian Bible. It has often been pointed out that ambiguity facilitates reinterpretation of a text over time, without the need to acknowledge that it has been altered. Spiritual language differs from scientific language of necessity. For spiritual texts breadth of suggestiveness increases apparent profundity while in science, fuzzy language undermines credibility. To examine use of language in texts whose function is spiritual and then assume that its vagueness makes it unsuitable for scientific thought misses the evident fact that languages have subsets with different vocabularies and rules for different sorts of use. Accordingly, we can dismiss the suggestion that the Chinese language is inherently unsuitable for formal logic or science.

A more detailed refutation of the once common Western misconception that the Chinese language was incapable of expressing abstract ideas is provided in Erbaugh (2002:51). In Chinese as in all languages, new forms of expression are easily invented when needed. Modern Chinese has no difficulty in devising terms for science and technology. Some are amusing, such the Cantonese "*diannao*" for computer, literally "electric brain" but the meaning is clear to all. As an example of how this occurs in English, consider the now pervasive use of acronyms such as IT, AIDS, MRI, as well as words borrowed from quite different contexts, such as hardware and software, which serve to resolve the pervasive need for precision in stating complex concepts quickly. Because this expedient is unavailable in Chinese, often the English acronym is transliterated, but in a way that may not sound similar to an Anglophone.

It must be admitted that in comparison to Aristotelian logic, Buddhist logic, both Indian and Chinese, is limited in that it

...cannot systematically and freely abstract from the concrete terms...to formulate the formal structures governing valid inference. [It] invariably discusses concrete examples, and moreover always examples that are of Buddhist doctrinal significance." (SCC VII:1:406)

We shall see something similar when considering Chinese science: there are many instances of careful observation, in astronomy, in medicine, in ecology, but these are not generalized into scientific laws. More often than not, they are fitted into correlative cosmology and left there. Thus both logical inference and empirical observation were developed in traditional China but tended not to progress to more inclusive abstract principles as does modern scientific thought. Of note also in Harbsmeier's statement is the close association of formal logic with Buddhist philosophy. Its value was in a spiritual context where it was applied to develop and defend rather abstract metaphysical doctrines regarding the nature of reality, perception, and mental processes. While the discourse was scholastic rather than empirical, it was intended to clarify difficult concepts and so bears some affinity to science. Thus the assumption that spiritual thought, in contrast to scientific thought, can only be fuzzy is not invariably correct.

Spiritual and religious language has considerable variety. It can be quite precise as in Buddhist logic or suggestive as in devotional or mythic literature. Religious concerns do not inevitably abolish precision of thought. Harbsmeier notes, however, that the logic imported from India as part of Buddhist philosophy did not hold much interest for the Chinese. Rather than take this as lack of a scientific attitude, it must be pointed out that Indian Buddhist logic is extremely abstract and, as Harbsmeier rightly points out, "essentially concerned with the justification of orthodox Buddhist claims...with promoting reasoned, rational discourse, but within this specific area..." (SCC VI:1:373). Reading Buddhist logic is challenging but its relation to scientific principles of deduction from data is essentially nonexistent. It must be pointed out that Western medieval logic was also developed to serve religion. The relation of science to formal logic should not be overstated. Science must follow logical principles such as the law of the excluded middle but logic alone cannot produce science.

There is a strong Chinese tradition, often likened to the paradoxes of Zeno, of exposing the limitations of purely verbal or logical thinking. Thus the first line of the *Dao De Jing*, one of the best known in all of Chinese philosophy, states, "The Way that can be spoken is not the unchanging Way." In a well-known anecdote Zhuangzi, as he was walking a river with a friend, remarked:

See how the minnows come out and dart around where they please! That's what fish really enjoy!"
    Hui Tsu said, "You're not a fish – how do you know what fish enjoy?"
    Chuang Tzu said, "You're not I, so how do you know I don't know what fish enjoy?" (Watson 1968:188–189)

Although the meaning of these has been in contention for over two thousand years, it is clear that they represent recognition of the limits of language and a statement of the value of subjective experience. It would be a mistake to regard Laozi or Zhuangzi as antirational. Rather they remind us that important aspects of human experience are subjective. This attitude, of course, is not empirical, but there is no reason to suppose Zhuangzi would oppose the methods of science, rather he would point out that there are aspects of mental life that science leaves out.

It has become a cliché that the West developed the objective aspects of thought while Asia developed subjectivity. This is a distortion since all cultures must develop both modes of thought. A more accurate way to put this is that China saw the universe in human terms, indeed, this is embodied in the tripartite formulation of heaven and earth with mankind in the center. Chinese philosophy always returns to the human meaning of the phenomena it treats. Yin and yang are as much aspects of immediate experience as they are abstract principles. In this sense, Chinese philosophy always stays close to what we could consider spiritual concerns, concerns that formal logic seems isolated from. Indian philosophy also, despite its vast durations and nearly limitless spaces, also is ultimately concerned with the human because it always assumes that correct understanding has redemptive value.

Needham (VII:2:90) quotes Nathan Sivin's formulation of a Chinese "law of inevitable succession" as follows: "Any maximum state of a variable is inherently unstable, and evokes the rise of its opposite." This repeats in modern terminology the fundamental idea of yin and yang to be found in the *Yi Jing*: that when yin or yang reach their maximum, they decline while the complementary force increases. Needham argues that the Chinese assumption that all objects and processes transform into their opposites made the rigid distinctions of formal logic, such as A and not -A, seem poor descriptions of natural processes. He also observes that formal logic has not been of much use to the actual conduct of science and so the Chinese lack of interest in this discipline was not an inhibitory factor in development of science. This accords with my own experience; even scientists capable of analyzing with extreme rigor, do not use formal logic and did not, so far as I know, study it in college. Formal logic remains an abstraction of limited use except in specialized fields such as computer science.

## ANALOGICAL AND SCIENTIFIC THINKING

Argument in Chinese philosophy is nearly always by means of analogy. Within formal logic, arguments from analogy are flawed because the two instances are never exactly alike. However analogy can have considerable rhetorical force and is universally employed in discussion and debate in all cultures. Poetry depends on analogy, which is after all, a form of metaphor, but as a mode of reasoning, analogy occurs in all kinds of discourse.

Yung Sik Kim (2000:306), in his study of Zhuxi, points out that while the philosopher's writings show considerable observation of the natural world, these were generally used for analogies to make ethical points about social behavior. His followers unfortunately limited their interest to the social message of these analogies, not at all in the natural phenomena on which they were based.

Here are some examples of Zhuxi's analogies based on the natural world, as summarized by Kim: "tigers manifest the father–son relationship, ants show the ruler–subject relationship; otters offer sacrifices to their ancestors; and *chiu-chiu* birds [osprey] show the proper distinction between husband and wife. As the basis of the last, Zhuxi wrote, "There are always a male and a female, and the two do not lose each other. Although the two do not lose each other, never is there a place, however, where they stand close to each other" (Kim 2000:194–196). Here the Confucian inhibition about male–female affection is readily apparent.

The osprey is significant because it is mentioned in the first line of the *Shi Jing*, the very ancient Classic of Poetry, a text filled with references to the natural world that later commentary transformed into moral exempla. In accord with this tradition, Zhuxi reduces the observation of the mating habits of the osprey to reinforce the Confucian dictum that men and women should be separate from each other. The interest in animal behavior is simply to make an ethical point. This is, of course, not much different from the representations of animals in moralizing folk tales of many cultures, including the fables of Aesop and LaFontaine.

The sixty-four hexagrams of the *Yi Jing* assume analogy between the sequence of broken and unbroken lines and any possible state of affairs. Numerology, too, is really a system of analogies between numbers and human situations. Thus for modern Cantonese eight is a lucky number because it is a homonym for "wealth" and because the character for happiness visually resembles the Arabic numeral "8". Similarly, 4 is unlucky because it is a homonym for death.

Analogy is common in science as a way of coining technical terms. To give a colorful example, the tiny depression at the base of the skull in which the pituitary sits is called the *sella turcica*, literally "Turkish saddle". However this is recognized as a poetic term for an anatomical structure, which is in no other way similar to a saddle. Indeed, the term is usually shortened to "sella" and many who use it are probably unfamiliar with its derivation. No correlation is drawn between the country or people of Turkey, or to horseback riding. Even the ubiquitous biological term "cell" was inspired by the resemblance of the cellular structures to the small rooms inhabited by monks which were termed cells. Here too, few who use the term are aware of its derivation and there no implication of real similarity of the life of a cell and a monk is imagined. With the analogies in Chinese philosophy, such as those of Zhuxi that we have just considered, things that are analogous are taken to have a real similarity; analogy as a form of argument seems not to have been questioned.

## DID THE CHINESE BELIEVE IN THEIR COSMOLOGY?

Despite the inhibitory factors we have been discussing, many scientific discoveries did arise in China. Nor was the traditional cosmology accepted uncritically by all. As early as the Han dynasty there were skeptics; by the Qing, much attention was being paid to the failure of orthodox cosmology to account for all astronomical phenomena. The views of Shao Yong, the Song philosopher who systematized Chinese numerology and advocated the number and symbol school of *Yi Jing* interpretation, particularly came under attack. Thus Wang Xishan pointed out that Shao's numerological use of four did not accurately describe the structure of the universe and that it ignored irregularities in transformative processes (Henderson 1984:187–190). Yet as Henderson points out, Wang had numerological systems of his own, for example wanting to divide the celestial sphere not by $365\frac{1}{2}$, the number of days in the year, but by 384, the number of lines in the sixty-four *Yi Jing* hexagrams. Despite the increasing recognition of the inadequacy of using numbers of spiritual significance to describe the natural world, numerology was not complete rejected. Numerology is greatly inhibitory of science because it impedes accurate mathematical description. Chinese astronomers were well aware that many phenomena did not fit the

cosmological scheme, but rather than stimulating a search for more inclusive theories of celestial phenomena, the response was to regard anomalies
are inherent in the cosmos, thus turning away from any effort to better account for them(Henderson 1984:250f).

Some scholars did recognize that seeking truth only in the classics conflicted with studying phenomena in themselves. As the Qing dynasty pro-
gressed, there was increasing interest in accurate mathematical astronomy,
to which those schooled only in the Confucian classics were unable to contribute. The presence of the Jesuits, who impressed the Chinese court with
the accuracy of their astronomical predictions, demonstrated the power
of empirical methods. This sparked the interest of many educated Chinese, though others denounced investigation of nature without ethical
purport as frivolous. Paradoxically, the Jesuits used science in the service
of proselytizing Roman Catholicism, hoping that their superior knowledge
would enhance the appeal of their religion. While several, notably Matteo
Ricci were greatly esteemed, they made relatively few converts among the
literati.

The syncretising tendency of Chinese thought allowed metaphysical
and empirical conceptions of the cosmos to exist side by side until the
superiority of Western science led many Chinese to reject the old ways of
thought as harmful to China. While the superior accuracy of astronomical
predictions was an important factor, politicians were most impressed with
the decisive power of Western armaments. The May Fourth Movement of
the early twentieth century advocated rejection of many of the old traditions, including the use of classical Chinese, and adaptation of Western
ways. Under Mao Zedong, though the official view of Confucius fluctu-
ated, it was the first time that outright rejection of Confucianism could
be official policy. Under the present more liberalized regime, study of the
classics has revived though they are no longer much taught in schools and
most Chinese have read only a few phrases of Confucius. A revived "Neo-
Confucianism" (Bresciani 2001), which is not to be confused with the Song
Neoconfucianism of Zhuxi, has perhaps more prestige than influence.

## ASIAN SCIENCE IN PERSPECTIVE

While science was most fully developed in Europe, it was never confined to Western cultures but arose in diverse civilizations as a result of
the inherent drive of the human mind to understand the strange universe
in which it finds itself. China did different things with its discoveries but
that is why we find its traditional culture of such interest today. What was
particularly lacking in China was a self-conscious conception of science as
a particular mode of knowledge with its own methodology. While there
were clear examples of empirical method, there was never recognition of

its immense power to create knowledge. Though criticized from time to time, the classics were still where one turned to fully understand reality. Empirical science never developed the continuity that was achieved by the spiritual-philosophical traditions of Confucianism, Daoism, and Buddhism. Each of these traditions was conscious of itself and aware of its origins, and each had a structure that was effective in transmitting its doctrines and traditions. Widely distributed canonical texts conserved their knowledge and sense of identity. Modern science has similar attributes though truth is sought in current journal reports that continually report new information, not in classical texts. We do not find anything comparable in Chinese or Indian science. Those we might consider scientists, or to use older terminology, natural philosophers, did not regard themselves as such and were not entirely aware of using a specific method that we regard as empirical science.

# Chapter 5

~

# The Metaphysics and Cosmology of Traditional India

Indian cosmology is extremely complex, consisting of many variant systems conflated with each other. Inconsistencies abound, as they do in Chinese cosmology which, however, was more unified than that of India. As we have already seen, inconsistencies are inevitable in correlative systems because myriad phenomena of the universe cannot be fitted into a single all-embracing system. Science as it develops often has conflicting theories but it is conscious of such inconsistencies and over time, reconciles them. In contrast, correlative systems tend to spawn inconsistencies as more and more is fitted into predefined categories and different systems are superimposed upon each other. In what follows, I will cover the predominant cosmological theories but this is not meant to imply that there were no variations on them. My discussion will be focused on philosophical aspects of Indian spirituality and is therefore an incomplete representation. I have not discussed *bhakti*, or emotional attachment to a particular god, despite its central importance in Indian life, simply because these practices have little relation to science. As in China, most probably focused on personal devotional practices and had but limited awareness of religious philosophy which was the province of religious specialists such as Brahman priests, Buddhist monks, and home-leavers in search of enlightenment. This emphasis on the ideas of the elite is unavoidable since science, with its dependence on education and leisure is an activity of the more affluent.

The most fundamental category of Hinduism was the four *varnas* (the correct word for caste, which literally means "color"): Brahmans, the priestly caste; Kshatriyas, warriors and politicians; Vaishyas, farmers and traders; and Shudras, servants (Smith 1994:9). Sometimes the last is left out so as to fit the number of varnas into the preferred number of three. Over the millennia, castes developed many subdivisions, which in some ways resemble tribes, with a complex array of clan-based animosities. It is likely that the castes rigorously maintained their supposed racial purity. Intermarriage, despite frequent social disapproval, has been pervasive in the lineage of *homo sapiens*. It remained social doctrine, however, that the varna distinctions were inherent in the structure of creation and within Hinduism there was no philosophical basis for abolishing them. The so-called heterodox traditions of Buddhism and Jainism however did not recognize caste distinctions. In practice however, followers of these religions in South Asia are not invariably free of caste prejudices.

The Brahmans were literate in Sanskrit, considered to be the perfect language in the sense of actually corresponding to reality; it was used for religious or other philosophical purposes. (Pali, closely related to Sanskrit is the language of the oldest Buddhist sutras. It is still in use by the Theravada sect of Buddhism in Thailand and Sri Lanka.) The Brahmans maintained the sacred literature which for centuries was transmitted orally from teacher to student. Education was not confined entirely to Brahmans; some Kshatriyas were active in philosophy and the historical Buddha, Shakyamuni was of the latter caste. As would be expected, not all Brahmans lived in accordance with their prescribed rules of religious purity, nor were all fully educated. Some were village fortune-tellers (Basham 1954:140). While less respected than pandits (scholars), it seems probable that their religious charisma as Brahmans enhanced their credibility as diviners and magicians. Jesuits were the first Westerners to learn Sanskrit, just as they were the first to master classical Chinese. In both cases the motive was to gain acceptance as religious teachers by speaking the languages associated with the ancient spiritual traditions.

Particularly in India, but also in China, correlative systems encompassed human society as well as the natural world. Social differences were assumed to be part of nature, not a creation of man. In the Hindu varna system social differences were fundamental:

The *varna* system managed to organize under one basic structure such seemingly diverse realms as the world of the gods, the division of space and time, spheres of what we would call the "natural world" (i.e. flora and fauna), and the realm of revelation and scripture. *Varna* in other words was a system which attempted to encompass within it all of the major sectors of the visible and invisible universe. But included within the *varna* system ... was the classification of society. (Smith 1999:8)

In China, society was also hierarchical but on the basis, at least in principle, of merit. There were hereditary class differences but these loosened considerably over the centuries; this did not occur in India. While in China, social position was rigidly hierarchical with detailed rules regulating permissible conduct at each level of society, high rank was available, at least in theory, by passing the civil service examinations, though there were restrictions on who was allowed to sit for these. Trade was despised, although by the Ming this was changing as the growing financial status of entrepreneurs in commerce forced society to open up for them. Otherwise, social mobility was mainly through the examinations in the classics in which the odds of success were not great and in which chance played a significant role in grading. Education was private and so a privileged birth into a wealthy family meant extensive tutoring with a greater chance, though not a guarantee, of examination success. Education was important in India but social position was determined by the varna of one's parents' rank so that social mobility was strictly constrained. Despite efforts at liberalization, caste distinctions are still powerful in India, though they are not easy for outsiders to detect.

## METAPHYSICS IN INDIA

Indian philosophers developed metaphysical theories of the nature of mind and objects that were quite different from those of China. While there was an interest in male and female principles, this was not a fundamental duality like yin and yang. Rather, sexuality was seen as an energizing principle pervasive in the universe. Both cultures had constrictive sexual moral rules; these were no more universally obeyed than in other cultures. Indian religion made use of much more explicit sexual imagery than did that of China; though erotic art was produced in both, it was less public in China.

While Chinese thought did become progressively more abstract and removed from ordinary experience, for example in the philosophies of Shao Yong and Zhuxi of the Song, Indian thinkers tended to more extreme levels of abstraction. Logical analysis (though of quite a different sort than that of Aristotle and the Western tradition), was often purely conceptual in contrast to China in which analogy was always the favored means of philosophical proof.

The Indian love of large numbers has already been mentioned. The mythology and cosmology presents near-infinite cosmic cycles and immense spaces populated with many thousands of supernatural beings. This reached an extreme in some of the Mahayana Buddhist sutras such as the *Avatamsaka* (Engl. *Flower Adornment*; Ch. *Huayen*). While both cultures conceived of time as cyclic, in India the cycles were inconceivably large. Here is A. L. Basham's (1954:323) description:

According to this system, the cosmos passes through cycles within cycles for all eternity. The basic cycle is called a *kalpa*, "a day of Brahma", or 4,320 million earthy years and his life lasts for 100 such years. His night is of equal length.... The largest cycle is therefore 311,040,000 million years long, after which the whole universe returns to the ineffable world-spirit, until another creator god is evolved.

There were also much longer durations referred to as mahakalpas. The exact numbers were, of course, fanciful, simply ways of expressing durations or spaces far beyond the capacity of the human mind to directly conceive. In comparison to a single human lifetime, durations were unimaginably long. A common phrase in Buddhism was that each of us has had as many past lives as there are grains of sand on the banks of the Ganges. Chinese cycles, in contrast tended to be on a human, or at least historical scale. Thus dynasties arose to replace degenerating ones, only to eventually degenerate themselves and be supplanted by yet others. The cycles of the *Yi Jing* and *wu xing* could be quite brief; the system for reckoning years was based on sixty-year cycles. The Indian notions of extremely long durations did enter China with Buddhism and were often referred to in the writings of Buddhist masters. Outside of Buddhist discourse, the shorter cycles of actual human experience predominated. Just as Indian mythology reached toward infinity, so its metaphysics sought to break events down into infinitesimal bits. This is most apparent in the notion of dharma in Buddhist metaphysics, to be explained shortly.

## THE MAJOR SACRED TEXTS OF INDIA

The canonical texts of India are quite different in character from those of China. Rather than being confined to the human scale, they are filled with myths and hymns associated with divine beings. The philosophical element is expressed less directly in the early texts, though purely philosophical treatises proliferated later. The Chinese classics, while they are certainly spiritual, have relatively little to say regarding deities. Confucius' famous statement that he respected the spirits but kept his distance from them is quite unlike the Indian attitude, which was one of great intimacy with their gods. There were popular gods in China whose images are still ubiquitous in Chinese temples and in households, which almost always contain an altar. But the philosophical role of Chinese gods was minimal. Proper sacrifice was necessary in both cultures to maintain the natural order, but beyond that the role of gods received different emphasis. Given the current tendency to see anything regarding supernatural beings as incompatible with science, it may seem tempting to regard the more impersonal official observances of China as closer to science than the more fantastical ones of India. Indeed, a prominent trend in early Western sinology was to imagine China to be an entirely rational society; this was largely a deist fantasy. There is no indication that these different religious attitudes affected the nature of science between the two cultures. Both made significant discoveries but neither developed an epistemology of empiricism.

The *Vedas* are the earliest sacred works to have been composed in India. Most scholars believe that they originated in the second millennium B.C.E. but were transmitted orally for centuries, at first because writing did not exist and later because they were thought too sacred to entrust to writing. As a result, attempts to reconstruct the development of these scriptures is a daunting task. Dating is uncertain because the exact time of formulation of oral material cannot be determined; tradition assigns the Vedas to an impossibly ancient mythical time. Even the early written material is mostly undated, in contrast to China in which precision in historical records was of greater concern. A usual explanation is that the Indian emphasis on enormous cosmic cycles and near-endless rebirth, together with the emphasis on transience, made time on the human scale seem unimportant.

The earliest of the Vedas is the *Rg Veda*, which contains cosmogonic myths, ritual instructions, and hymns and is thought to have been composed about 1200 B.C.E. An interesting portion of the Rg, concerns *Soma*, a plant of sacred significance whose juice produced an altered state and, supposedly, immortality. It was suitable for gods as well as humans and was used by the warrior god Indra (Brockington 1996:99). Soma was the focus of numerous rituals and itself became a god. If Soma was an actual plant, its identity was lost early. The importance of Soma in this earliest Indian sacred text fits with the hypothesis, much beloved of the contemporary drug culture, that mind-altering drugs may have been used early in human history to induce ecstatic religious experiences and that such experiences may be the origin of religion (Hayden 2003:284–292). This hypothesis is rather suspect due to the lack of historical evidence and due to its exaggeration by psychedelic drug proponents (Redmond 2004). In any case, in India as in China, religious states came to be induced by rituals and practices such as various forms of yoga (Brockington 1997:99). Later, in both Tantra and Daoism, sexual content became prominent as a means of inducing intense emotional states, though the extent that tantric sexual acts were actually carried out needs more complete study. Important here is that while pharmacology may have been involved in some religious activities, it played no part in the great majority.

Of more direct interest with regard to the development of science, with Soma we have one of the first records of human use of pharmacologically active plants. One of the great mysteries of the history of medicine is how humans came up with the idea that ingesting plants and other substances would have altered the progress of disease. It can be speculated that plants eaten as food were discovered accidentally to have effects on the body and brain—though not always favorable ones—and that this spawned the

88 Science and Asian Spiritual Traditions

hope that they could be used for healing. What is extremely puzzling is why, despite most plants being inactive pharmacologically, an immense variety came to be used in both India and China for medical use. Herbal formularies of both regions list many hundreds of plants with the basis for inclusion being difficult to determine. While many were added over the centuries, there does not seem to be an equal interest in dropping those already included on the grounds of lack of efficacy or adverse effects. This indicates that the herbal pharmacopeias functioned more like scriptures than scientific treatises. A scripture is a text believed to be based on the teachings of a god or religiously inspirited individual; given such author-ity, there is no justification for deleting material. With science, in contrast, correction is continuous. In spite of the current sentimental fondness for "natural" remedies, most of those in ancient formularies do not produce significant actions on the body and some may have serious toxicity. Yet medicine-taking is one of the most persistent features of human culture.

The *Atharva Veda* was the last of the Vedas to be composed and is some-what different in character from the other three. Like the earlier Vedas it contains many hymns, but also much of what might be considered prac-tical content. Healing methods are described, both exorcistic and herbal. Fasting is also recommended as a healing modality, a practice that has continued to be prominent in Indian culture for both spiritual and medical purposes. Traditionally, the *Atharva Veda* is considered the foundational text for Indian medicine. In addition to the Vedas, there were a large number of subsidiary texts, some of which treat of science and medicine.

The origin of science and technology in India is associated mythologi-cally with gods, as described in the Vedas and other sacred texts. In China on the other hand, these beginnings are attributed to quasi-human mythi-cal culture heroes such as Fu Xi, Yao, and Shun. In the history of religion, there is not always a clear difference between mythical humans and gods. The Buddha is a salient example, being regarded as an exceptional human in some contexts and as supernatural in others. Whether origins are as-signed to gods or to mythical humans has had no consistent effect on the development of science. It can even be argued that the association of sci-ence or technology with the divine tends to make its findings more likely to be valued and thus preserved.

## THE METAPHYSICS AND PSYCHOLOGY OF CHANGE

In Indian metaphysics all phenomena pass through a three-part se-quence of arising, enduring (or persisting), and ceasing (passing away). Sometimes the final phase is divided into two stages: waning and extinc-tion, but most often only three phases are mentioned. These phases seem closer to common sense than the wu xing (five phases) of China. They function in mythology, in metaphysics, and in meditation. There is a close relation of this sequence to the experience of *vipassana* or insight meditation, the only form that seems to have been taught by the historical Buddha, in which one closely examines one's own thoughts. Practicing vipassana results in comprehension on the deepest level that all mental phenomena are impermanent: arising, persisting, and then passing away. The goal is to develop a subtle awareness of the truth of transience, referred to as *anicca* in Buddhism. Since the mind and reality are ultimately identical, this form of meditation also brings insight into the actual nature of phenomena, which like thoughts, arise, persist, and pass away. With meditation this insight is experiential, not merely verbal. Recognizing the true nature of reality leads to release from the unsatisfactory nature of ordinary human life and, metaphysically, from the repeated rounds of birth and death. I have described this in Buddhist terms; in Hinduism, one recognizes Atman, the ultimate, unchanging principle underlying all phenomena. While the Hindu doctrine of Atman, or a permanent, cosmic self and the Buddhist one of Anatma, or no permanent self sound opposite, there are many similarities in actual meditative practice and in other doctrines of the metaphysics. Neither regards what is experienced by the senses as ultimately real. As noted in the preceding chapter, Joseph Needham regarded this viewpoint as antiscientific. The contrary argument is also possible: by directing attention beyond the flux of momentary events, general principles can be discovered. Related arguments have been made by Capra and others (1991). That opposite arguments can be made regarding the same philosophical outlook suggests that the specific content of these metaphysical systems does not have any immediate bearing on the development of science. However, as previously pointed out, metaphysics is a quite different style of thought than science and the two do not really mix.

These three phases of phenomena are not only philosophical; they are represented in Hindu mythology as the three most important post-Vedic gods Brahma, Vishnu, and Shiva who are, respectively, the creator, the preserver or maintainer, and the destroyer. Destruction is not negative in this context however for two reasons. First, and most obviously, the old passes away to make room for the new. Secondly, cessation ultimately refers to cessation of the limitations of the human state. That all phenomena eventually come to an end means that suffering can end also. The end of craving and delusion is enlightenment, a possibly psychological state but also the metaphysical state of nirvana, which is beyond the merely human. Thus, because it does not give way to a new round of birth-and-death, with its inevitable pain, cessation is to be celebrated. Though Hinduism and Buddhism develop these doctrines is different ways, both agree on the central importance of ending the clinging to delusions that is the cause of suffering.

## INFINITESIMAL METAPHYSICS AND MEDITATION

For most moderns, the most interesting aspect of Indian metaphysics is the way that analysis of the ultimate constituents of phenomena is integrated with the systems of meditation intended to bring release from suffering. In vipassana meditation, and its related early Buddhist metaphysical system of *Abhidhamma* (Bodhi 2000) the phenomena of the mind are broken down into infinitesimal bits called *dhammas* in Pali. The Sanskrit form is *dharma,* not to be confused with upper case *Dharma,* meaning law or teaching. I use the Pali form here because the detailed theory was written in that language. The dhamma theory does break both mind and matter into tiny fragments. While this is closer to infinitesimal calculus and to particle physics than the Chinese doctrines of yin-yang and wu xing, the similarity is limited. While it is true that subatomic particles go through the three phases of arising, enduring, and passing out of existence, the metaphysics of dhammas, while it breaks reality down to sub-elements, is intended to analyze thought, not objective matter and energy. Yet Abhidhamma is a highly dynamic mode of analysis and perhaps an easier fit with modern physics than the yin-yang and wu xing interactions of Chinese metaphysics. As all who have studied the history of science know, the Greek philosopher Democritus analyzed phenomena into atoms. Greek atoms and Indian dhammas were both metaphysical, not empirical, and neither led directly to the discovery of particle physics. What these ancient concepts do indicate is the tendency of the mind to break reality down into very small subunits. This analytic tendency of the human mind is part of what makes science possible, so in this sense we can say the Abhidhamma and Democritus' atomic metaphysics show the spirit that many centuries later led to science.

Abhidhamma purports to account for how the mind works and so is a form of psychology, but with the metaphysical assumption that mental and external reality are the same. This is not so simplistic as to maintain that whatever we want to think about the world can be true. Rather it assumes that both thought, and the apparent external reality that it perceives, are equally illusory. The experience of vipassana meditation is assumed to confirm the Abhidhammic analysis. Many who practice vipassana come to feel that it is a useful way to better understand some aspects of their own minds. However, evidence for the benefits of vipassana practice is no less subjective than the evidence for any other from of nonexperimental psychology. Even if meditation does seem to confirm the truth of the cycle of arising, enduring, and passing away, this is, at least in part, because the meditator begins this sort of meditation with the expectation that his or her practice will confirm this metaphysics.

It is far from clear how the goal of meditation, understanding of one's own mind, might be objectively confirmed. From a therapeutic point of view, systems of psychotherapy should be confirmed based on objective evidence of whether patients benefit from them. Such evidence has been elusive because it is impossible to control the many variables involved. For Abhidhamma and vipassana, we might say they are as close to being scientific as many contemporary forms of psychological treatment, but no more. In traditional India and China, meditation was not regarded as a form of psychotherapy; the concept did not exist. The much stronger claim was made that insight can lead to release from the human state and all its suffering and limitations. This claim, though it motivates many modern practitioners, cannot be tested empirically. There are however objective brain changes associated with meditation; research on these continues (Austin 1998). Such research must be interpreted with caution because it cannot be assumed that changes in hormone levels or brain waves are somehow the basis of any claimed religious effects of meditation.

Abhidhamma is cognitive—it proposes to end our suffering by more accurately recognizing how our minds process information. Centuries later, both Hinduism and Buddhism developed systems of belief and practice known as *Tantra*. These originated in India and have survived in Tibet. While Tantra has cognitive components, it emphasizes emotional imagery, often of violence and sexuality, to startle the practitioner into enlightenment. Tantra has an avid following in the West, aided by its sensational sexual imagery as well as the charisma of the Dalai Lama. The religious justification for use of such visual melodrama is that it aids in using the energy of intense emotion to push oneself toward realization. Images of gods and goddesses in yab-yum (sexual intercourse) evoke sexual energy. Many tantric practices often involve frightening forms of deities with destructive powers. To deepen understanding of impermanence, meditation amidst rotting corpses in charnel grounds is recommended, if not often practiced. At the root of the destructive images of deities is the idea of cessation so that Tantra can be seen as augmenting the rather dryly analytic method of Abhidhamma with imagery that uses emotion rather than suppressing it. The destructive forms of gods also represent the destruction of erroneous views that is essential if one is to achieve enlightenment. For this reason, such practices have sometimes been likened to some modern forms of psychotherapy, fortunately no longer in fashion, that attempt to provide catharsis of complexes by releasing early traumas. However with both Hindu and Buddhist practices, harmful ways of thinking are not interpreted as resulting from individual traumas but are inherent in the sentient state. The method for attaining release from suffering is set by the sect to which one belongs; recollection and verbalization of recollection and analysis of early experience are not involved. Indeed, blaming one's parents for one's neuroses would be contrary to the emphasis on filial piety in both India and China. There were systems for classifying personality types in both India and China but these consisted of a few categories, there was nothing comparable to the elaborate analysis of individuals' lives found in Freud and many later theorists of psychotherapy. Nonetheless, we can see broad similarities between ancient and modern psychological theories. Abhidhamma and vipassana are akin to cognitive psychology in that they emphasize more realistic ways of thinking, while Tantra has some resemblance to Freudian and other forms that focus on sexuality. It is interesting that such distinctive styles of psychology are quite ancient. As to how effective they are, the jury is still out more than two thousand years later.

## THE THREE GUNAS—THE FUNDAMENTAL FORCES

Dhammas are the constituents of mind and matter; each goes through the three stages of arising, enduring, and passing away. A separate theory, the three *gunas*, described the forces that mediated change. Like dhammas and the three stages of phenomena, the gunas act on multiple levels (Desai 1997:670). On the cosmic level they are forces: *adana* (centripetal-attractive), *utkranti* (centrifugal or inertial), and *pratistha* (equilibrium). These also act on the personal level. Here the centripetal force is *sattva* or being; it holds the constituents of existence together. Centrifugal force is the tendency of things to fly apart; on the human level it is called *tamas*. It corresponds to darkness, dissolution, and the dispersion of individuality. Like cessation, tamas is ambiguous representing both death and release from limitation. *Rajas* is the revolving tendency and the basis of all motion. These are also the three stages of phenomena as previously described: rajas is creation or Brahma, sattva is preservation or Vishnu, and tamas is destruction or Shiva (Danielou 1964:22–24). In their association with gods, the gunas also correlate with parts of the body: Brahma, the heart, Vishnu, the navel (considered the seat of consciousness), and Shiva the sexual areas. It is easier to relate the gunas to the forces of physics than the five phases of Chinese cosmology. Yet the correlative associations of the three gunas cannot be compared to anything in modern physics.

The devout believer may be thinking of anthropomorphic traits rather than metaphysical ones while performing devotional practices to Vishnu or Shiva. (Brahma is not often the object of devotion.) Indeed, viewing the sinuous and elegant Chola period bronzes of Shiva beside his voluptuous consort, Parvati, hardly gives rise to philosophical musings about destruction. Similarly, Shiva as lord of the dance represents not only the play of illusion, but also the sensuous rhythm of creation. Here as is often the case, a devotional form of a deity has somewhat separated from its abstractly philosophical attributes.

The three gunas are fitted into the correlative cosmology on all levels. Each corresponds to a state of consciousness: rajas is wakefulness or awareness, sattva is dreaming, and tamas, dreamless sleep. These in turn are associated with different kinds of meditation. Gunas are also combined to generate more hierarchical categories. On top is sattva in sattva, the ultimate, unchanging self; the lowest, tamas in tamas, is the inanimate world (Danielou 1964:27f). Each guna correlates with ethical and character traits and also with vulnerability to different diseases, based on which guna predominates in his or her constitution.

### The Gunas and Science

In explaining the gunas with terms borrowed from physics such as centrifugal and centripetal force and inertia, I am following common usage of Indologists who employ these analogies to explain these extremely abstract entities (Danielou 1964:22, Desai 1997:670) This does not mean that the gunas prefigure the forces defined by physics. Thus rajas as the cohesive tendency and tamas as the dispersive correspond only approximately to the forces of physics. Gravity and electromagnetic forces can draw objects together, or in the latter case push them apart depending on polarity, but Indian writings do not refer to anything like gravity or electromagnetism. Rather the gunas represent a way the mind organizes disparate phenomena based on their human significance.

## THE INDIAN MODEL OF THE UNIVERSE

Indian religions developed very detailed models of the universe divided, like most other such systems, into three levels. Different regions of the virtual geography correlated with different spiritual states. In early Buddhism the world was envisioned as consisting of concentric circles with Mount Meru at the center and the world that includes humans, *Jambudvipa*, in the south of an outer circle (Sadakata 1997:25–40). The sun, moon, and stars are in an outer belt of wind. Below these circles are the hells, populated with demons and evil souls awaiting their next incarnation. Above are pleasurable realms but these are not like the Christian heaven. In the Indian systems, even the highest divine beings in the heaven will, once their immense stock of favorable karma has been used up, be reborn into the round of birth-and-death (samsara) (Sadakata 1997:41–68). Humans are actually better off than the gods because there is just enough suffering in the human realm to motivate them to attain permanent release from samsara by following the Buddhist path. The Buddhas themselves have permanently escaped samsara and are in nirvana, an ineffable state that is not given a location within the universe. The Mahayana, which developed later, followed the Indian tendency for progressive elaboration; several of its most influential sutras, such as the *Saddharma Pundarika* (Lotus) and the *Avatamsaka* (Flower Ornament) (Cleary 1993), describe vast Buddha-fields in which a cosmic Buddha preaches to Bodhisattvas and innumerable other divine beings. Like the parallel universes of the modern philosopher David Lewis (Lewis 1986), these Buddha-fields have no particular location in relation to the ordinary world in which we live. Unlike those of Lewis, we can know about them but how such knowledge is possible is never really explained.

The cosmology presented in these Mahayana sutras bears no significant resemblance to the universe as understood by science. This is in contrast to that of Ptolemy whose account was based on data, though his geocentric model was eventually refuted by more refined observation and better geometrical insight. The universes of Buddhism are imaginative creations, developed to support a religious vision of human life. In the change from the early Buddhist Jambudvipa model to the vast spaces of the Mahayana, we can see the emergence of a sense of the measureless extent of the universe, a desire to extend thought beyond the reach of what can be observed, perhaps associated with the expansion of Buddhism far beyond its northeast Indian homeland. While this cosmology did not directly stimulate the development of astronomy in India, coming to conceive space as near infinite suggests the expansion of human interest beyond the everyday and close at hand.

## DID THE INDIANS BELIEVE IN THEIR OWN COSMOLOGY?

The same question was asked in the previous chapter in regard to whether the Chinese took seriously such notions as yin-yang that now seem at best metaphorical. India too had its skeptics. Indeed the inhibitory effect of religious doctrine on the development of Indian science was noted more than one thousand years ago. One of our main sources for Indian science is the writings of the Moslem, Al-Biruni, also transliterated Al-Beruni (973–1051 C.E.). Among his scientific treatises were one on astrology, in which accurate observational measurement was emphasized, as well as a volume on mineralology and a herbal formulary (Samian 1997:157f). Although he wrote from outside Hinduism his insights are valuable. Here is his statement about science in Hindu India:

... even the so-called Scientific Theorems of the Hindus are in a state of utter confusion ... always mixed up with the silly notions of the crowd, e.g. immense numbers, enormous spaces of time, and all kinds of religious dogma. ... I can only compose their mathematical and astronomical literature . . . to be a mixture of pearl shells and sour dates, or of pearls and dung. . . . Both kinds of things are equal in their eyes, since they cannot raise themselves to the method of . . . deduction. (Rahman 1998:16)

Given that Al-Biruni was a Moslem, his view of Hindu science was perhaps not dispassionate. Despite its unsympathetic bluntness, Al-Biruni's view of the science of traditional Asia is apt, at least from the point of view of the scientist, who finds impressive discoveries diluted with myth and metaphysics. The historian of religion however would not follow Al-Biruni in dismissing the nonscientific parts as "dung" but rather be open to considering what value these aspects had to Hindus. Yet the cultural value of correlative cosmology cannot counteract its inhibiting effect on science. Rahman summarizes this point well:

Actually, this phenomenon is common to all cultures in every period. The political and social pressures exerted by the people . . . are exploited to dilute the content of science or to attack science whenever its conclusions come in conflict with the prevailing philosophy, the political interests or the social and cultural features of the elite. This is very much in evidence in contemporary society, including those of the so-called scientifically advanced countries. (Rahman 1998:17)

This serves to remind us that there are always factors outside of science that limit the acceptance of some of its findings. Some, but by no means all, of these factors are religious.

# Chapter 6

Astronomy and Astrology in
China and India

## THE NATURE OF TIME

As a physical measurement, time is identical in all cultures; as a component of human experience however there are differences. Confucius pointed to the fundamentality of time an oft-quoted phrase, "Look at the river, endlessly flowing." Here the word "look" is significant because rather than conceiving of time as an abstract process, the Chinese focused on it as experienced by humans. This is quite unlike the near-infinite durations of Indian cosmology. We are considering here philosophic aspects of time; the length of a day, a lunar month, or a year was the same everywhere.

Most distinctive about time in Chinese culture is its qualitative character, expressed concisely by the great Han Chinese historian Sima Qian (145–186 C.E.):

What I call time is not the passage of time. Men inherently have propitious and nonpropitious times (Harbsmeier 1995:52).

Thus time was of interest as a factor in human affairs. It was inherently spiritual in the sense that it determined what one's life was like at particular intervals. Not only the *Yi Jing*, but almanacs and an immense variety of divinatory methods were used to assess what actions were favored at a particular time. One, still in use in Hong Kong, was to have a caged bird select from a pile of paper slips, each containing a fortune. The variety of such methods attests to the Chinese belief in an underlying, but mostly hidden, natural pattern that humans needed to discover in order to be successful in any undertaking.

This is not to suggest that the Chinese did not also employ objective measures of time. The most common device used was the sundial but other devices included the clepsydra (water clock) with measuring sticks divided into one hundred segments. Incense sticks were another means, approximate but preferred for timing meditation sessions in temples because they also served as offerings to ancestors, or the Buddha. Time could thus be determined and stated with a level of precision that was quite adequate in an era when there were no time clocks to punch nor planes to catch. The most widely used system for designating time was division of the day into twelve segments named after the same zodiac animals used to designate years: rat (11 P.M. to 1 A.M.), ox (1 A.M. to 3 A.M.), tiger (3 A.M. to 5 A.M.), etc. This system of names rather than numbers lent itself to expressing the qualitative nature of each segment of the day. The time of day also had a yin-yang correlation. Midnight, for example, marked the transition from the yang phase that began at noon to the yin phase that, in turn, ended at noon. In the contemporary world also, each time has an affective quality and rules as to which actions are appropriate or inappropriate. Thus for many six or seven A.M. means getting up for work, five P.M. or later means the end of the work day and the onset of commuting, followed by domestic tasks and leisure. Coffee is taken in the morning but not alcohol. Similarly, days of the week are qualitatively different. Most feel far happier on Friday than Monday, for example.

An interesting sidelight on the spiritualization of time in China is the deification of the Jesuit Matteo Ricci, who introduced the self-chiming clock to China in 1601. On his death, this Catholic missionary became the god of clocks and, in Buddhism, the Bodhisattva Limadou (Wilkinson 1998:222). Here we have an example, far from an isolated instance, of technology inspiring religion. A devout Catholic, Ricci could not have been happy if he could have known of these posthumous distinctions. Religion abounds with instances in which people, events, and objects are appropriated into systems of meaning quite unlike their original ones. Though the Chinese were quite capable of understanding how clocks worked, this did not exclude the necessity of having gods to watch over their functioning. As we shall see in the cases of smallpox vaccination and ceramic manufacture, technological processes tended to have their own dedicated deities.

The Kangxi emperor, who held Ricci in high regard, established a factory for manufacture of mechanical clocks. Later, under the reign of the Jiaqing emperor in 1796 the import and manufacture of mechanical timepieces were banned, though with doubtful effectiveness. Thus the mechanical clock's spiritual journey in China began as a means to impress the emperor's court with the superiority of Christian culture, where it quickly became an imperial plaything, then had its own deity and Bodhisattva dedicated to it, only to be transiently banned as foreign and corrupting.

This reminds us that the religious and political significance of science and technology often shifts over time. In our time too, scientific research can serve as foci for cultural concerns, two examples are nuclear energy and embryonic cell research.

It is often said that in the West, time is conceived of as linear while in Asia it is cyclical. The basis of linear, irreversible time in the West is both religious and scientific. In Christianity, the central event is the birth of Christ as the direct entry of God into the world. The salvation offered by Christianity is possible only for those living after this event. The implication of this doctrine for time measurement is apparent in the convention of writing dates as B.C. and A.D. For the Christian, the human situation is absolutely transformed by the birth of Christ; the years B.C. and A.D. are entirely different with respect to the possibility of human salvation. Perhaps related, since it began in the Christian West, is the notion of progress. The assumption that knowledge becomes progressively more complete is inherent in science and has been prominent in Western political theory. The so-called Whig theory of history interprets history as progressive improvement in society toward greater freedom and greater concern for human welfare. In science we have the irreversible big bang and evolution; in politics, liberal democracy and, in a quite different way, Marxism. In traditional cultures, the theory of history is almost the opposite. A fundamental myth is that the earliest time of human civilization is taken as a time when life was simpler, but peaceful and happy. The Garden of Eden myth is but one form of this. Thus improvement of human society is conceived of moving backward in time, not forward.

The distinction between a Western conception of time as linear and an Asian cyclical one is not absolute but one of emphasis. Both civilizations were aware of history as a series of unique events and both saw certain aspects of life as cyclical: the seasons, the movements of the heavens and, most fundamentally, the nature of human life from birth, to work and fertility, to old age, and ultimately to death. In China, as in any agricultural society the seasons determined what activities were appropriate. For the elite, who were spared agricultural labor, there was more freedom in allocation of time but therefore more need to make decisions about it. The organic notion of time was particularly developed in China and is exemplified in the classic divination manual, the *Yi Jing*. The separate oracles in the Changes virtually all concern what activities are favorable at the moment of consultation, expressed in recurrent phrases such as "favorable to cross the great river" and "favorable to meet the great man." Crossing the great river refers to any major task—in the era before suspension bridges, crossing a raging river in a tiny boat was not to be undertaken lightly. Using the *Yi Jing* tells what sort of human actions are likely to succeed in the near future, and which are not. This aspect of time is real; we all take care to decide the right moment for such important actions as asking the boss for a raise or proposing marriage—or breaking up. As experienced in our lives, time is not an abstract entity, a variable in a physics equation, but something of practical and affective significance. The *Yi Jing* and astrology are, to use modern jargon, guides to time management. The need to make decisions about timing accounts for much of divination's persistent appeal, despite its unscientific nature. The time of physics is not the time of human lives but that of the *Yi Jing* and astrology is. This is an instance in which science does not meet a human need, causing many to turn to methods that seem spiritual. While modern reliance on divination is perhaps not fully rational, millions turn to it to gain confidence regarding difficult decisions. Viewed in this perspective, it is apparent why science has not led to abandonment of divination: it does not provide an alternative means of understanding time. The modern world does provide other ways of aiding in decision making but many, from weather forecasts to investment advice are far from infallible.

In India, time was human also but in a quite different way because the doctrine of rebirth made each sentient being's existence almost endless. Philosophically at least, one was concerned about the very long term. Indian sacred texts are filled with references to unimaginably long cycles measured in kalpas and mahakalpas, defined in fanciful fashion. Though most definitions, as discussed in an earlier chapter are quite abstract, some use a human point of reference. For example a common definition of a kalpa in Buddhist texts is the time it takes to wear down a large boulder if it is brushed by a piece of cloth or a feather once every one hundred years (Soka Gakkai 2002:325–330) This is of course a very long period of time but it is described with a metaphor that is easier to picture in the mind than most in Indian texts.

Indian cosmological literature describes four types of kalpas corresponding to the cycle of arising, continuing, waning, and ceasing. During this long cycling, the human lifespan increases to eighty thousand years, then gradually diminishes to ten years. As this indicates, time in India was an object of metaphysical speculation, conceived in transhuman terms. Though Indian ideas of time traveled to China with Buddhism, these were not seen as contradicting the human time of the *Yi Jing*. Nor need they have been since time is both infinite and momentary. Human and cosmic conceptions of time are not contradictory but rather emphasize two distinct aspects of the mystery of time: the abstract and infinite and the immediate and human. We experience time's effects but can represent it only by metaphor, as in the quotation from Confucius that began this chapter.

Indian cosmology has been likened to that of modern astronomy in its near-infinite duration, while the relativity of time in the *Yi Jing* has been compared to Einstein's theory of relativity. This resemblance is very limited. As we have seen, long durations in India were defined in metaphorical terms and intermixed with notions, such as the change in the human life expectancy, that are entirely mythological. The relativity of time in the *Yi Jing* has to do with its subjectivity at different times of a person's life; that of Einstein refers to physical parameters such as the speed of light.

## ASTRONOMY BEFORE URBANIZATION

Because the night sky is inherently mysterious and remote, it evokes a sense of the numinous and implies realities beyond those of everyday life. Astronomy thus is among the earliest subjects of human speculation.

For urban or suburb dwellers, viewing the stars and planets requires deliberate effort. Indeed, most are probably more familiar with celestial objects as photographs in a science magazine or planetarium shows than from actually looking up at the sky. However, a night walk in the remote country or on a desert reminds us how dramatic is the appearance of the night sky and puts us in a position to appreciate how profound an effect the sight must have had on people who had no conception of the physical nature of heavenly objects. Much attention must have been lavished on the sky if only for the simple reason that at night there was not much else to look at. The positions of sun, moon, and stars changed over the course of days and months, but in repeating patterns. Observation of these regular patterns, different for sun, moon, planets, and stars, must have been a factor in the universal belief in the underlying regularity of nature, of which humans were a part. Planetary motion is more complex than that of stars and for this reason attracted particular interest. Anomalous events such as meteors, eclipses, and comets must have been conspicuous to those accustomed to observing the sky every night and gave rise to fears that the natural order was threatened.

It must be emphasized that the role of celestial events in regulating human life was not imaginary. In societies dependent on agriculture for survival, the movements of the sun largely determined the patterns of human life by creating daily cycles of light and dark and yearly change of the seasons. Understanding the seasons was a matter of life and death when adequacy of food supply depended on the success of each year's harvest. The rich could store up food but also depended on laborers who had to be fed.

Given that seasons were the most dramatic and also predictable of all patterns, it was natural that celestial events would be assumed to control other aspects of life. Because the celestial macrocosm was believed to respond to human behavior, unexpected celestial events, such as comets and eclipses, could be interpreted as the judgment of heaven that the emperor's behavior was not exemplary, though it was usually neglect of ritual observance rather than governmental policies that were thought to be the cause of heaven's displeasure. For this reason, astronomical events could threaten political stability. A class of imperial advisors owed their employment to their supposed ability to interpret such portents. Needless to say, their interpretations served their political agendas, most often to curry favor with the emperor.

The analogy of celestial patterns with those of human life must have arisen very early in the development of human consciousness. Their regularity was reassuring, just as unexpected changes were frightening. If one were to condense Chinese philosophical preoccupation to a single concern, it would be according with the natural and social order in the midst of change. This was not simply an abstract conundrum such as how we can call the Yangzi River the same river since the water we see flowing through it is always different than it was a moment before. Rather it was how one can best live in the face of constantly changing circumstances.

Anomalous events of any kind exacerbated what was the most persistent anxiety in Chinese civilization: loss of order. Underlying the philosophical analysis of change was fear. Thus the primary concern of all the major philosophers was analyzing change and teaching how to maintain the natural order. While there was faith that the natural order would be restored eventually in all circumstances, there was great danger before this restoration occurred.

The *Dao De Jing* gives an account of permanence and change in terms of the relation of the unchanging Dao and the ten thousand, that is, myriad things. Buddhism teaches that impermanence (*anicca*) is inherent in sentient existence and offers meditative practices to overcome its painful effects. Confucius in the *Lunyu* often advises how the superior person should behave when political circumstances are unfavorable. The assumption was that correct ritual propriety would, by according with the mandate of heaven, maintain social order. This seems peculiar now when we regard social policy as the key to effective government. In actuality, the Chinese imperial system of government was dysfunctional. On the one hand, during much of its history the system did maintain order, though life for the nonelite was miserable, as was generally true in the ancient world. On the other hand, government control over such a vast territory in an era without rapid communication was always difficult. In practice local officials were frequently corrupt and cruel and the people had little recourse to the central government. Confucianism saw the solution to loss of order as more correct personal and ritual behavior to try to win back the mandate of heaven. There were pragmatic officials who did their best to aid the common people, but Confucian principles offered little practical guidance. Nor was the use of divination to determine courses of action well-suited as an aid to government policy, though some must have used divination as a way to lend authority to proposals actually inspired by common sense.

I have made these points because there is a tendency among some moderns to sentimentalize life under correlative systems in which each had a definite place in the cosmos, in contrast to the angst pervasive in the modern West. Correlative systems do describe the cosmos as ordered and did provide some mental security but did not facilitate management of disorder when it did occur. Here science is far more effective. By explaining the mechanism of eclipses, comets, and other seeming anomalies, it removes anxiety about these events which now can be seen as part of the natural order. With the advance of astronomy during the Qing, prediction of eclipses improved and their destabilizing potential was reduced. Similarly, science understands weather and earthquakes, though prediction remains imperfect. Governments now respond to such events with efforts to aid victims rather than with rituals. Unfortunately, disaster relief is often impeded by political factors.

## THE EARLIEST EVIDENCE OF CHINESE ASTRONOMY

Systematic astronomical observation in China was present even in Neolithic times because graves were sited to align with celestial bodies. Stars and asterisms, which can still be identified, were referenced to the oracle bones inscriptions, the earliest surviving Chinese texts and are recorded in later first millennium B.C.E. texts (Pankenier 2005:18–27). Thus the earliest astronomy was motivated by the belief that human actions had to be coordinated with astronomical events.

Important astronomical references in the *Zhou Yi*, the earliest layer of the *Yi Jing* have only recently been decoded, apparently because of the tendency of scholars to focus only on the literary and historical aspects of a text, without awareness of the scientific knowledge of the time it was composed. A brief examination of the history of one such celestial reference demonstrates the importance of appreciating science as an element of early cultures.

The top (final) line of the first hexagram, *Qian*, in the literal translation of Rutt reads:
See dragons without heads. AUSPICIOUS (Rutt 1996:224).

This phrase has baffled translators and commentators since it is far from obvious why dragons without heads would be auspicious. Legge translates it as follows :

If the host of dragons . . . were to divest themselves of their heads, there would be good fortune. (Legge 1899:58)

This is an expansion of the original Chinese phrase but does reduce its obscurity.

The most familiar English translation, that of Wilhelm and Baynes makes the phrase seem clearer but at the price of straying from the original.

There appears a flight of dragons without heads.

Good fortune.

Wilhelm glosses this to mean:

Strength is indicated by the flight of dragons, mildness by the fact that their heads are hidden (Wilhelm 1967:7).

Neither Legge's paraphrase nor Wilhelm's suggested interpretation is entirely convincing; both seem to be guessing at the meaning. Yet, since both had lived for many years in China and utilized Chinese informants in making their translations, we must assume that the original meaning had been forgotten in China by the late Qing. What is almost certainly the correct decipherment of the passage was achieved by the American scholar E. L. Shaughnessy (1983:268–278), who recognized that the dragon in question was actually an asterism whose head was not visible at the time of year referred to. He further establishes that the other five-line texts are references to seasonal positions of asterisms. Further evidence for the identity of the dragon as a constellation is line 5:

A dragon through the heavens glides (Rutt 1996:224).

If "dragons without heads" refers to a time of the year, expressed as the position of an asterism, then its designation as auspicious makes complete sense.

Richard Kunst disputes this interpretation, suggesting that the lines might have originally referred to the dragon as a mythical beast that were later (but still in the early Zhou) rearranged to correlate with positions of the dragon constellation (Kunst 1985:387–388). For our purposes it really matters little whether this entirely speculative suggestion is correct or not because in either case the text indicates careful observation of the heavens during the early Western Zhou. Though astronomical references are infrequent in the *Zhou Yi*, even allowing for the possibility that some have been overlooked due to the cryptic nature of the text, the single reference is sufficient to show that the early Chinese were aware of the regularity of seasonal changes in star positions. Unfortunately, this intriguing text cannot tell us how extensive was observational knowledge at the time of the *Zhou Yi's* composition.

This instance demonstrates that from earliest time, empirical observations served a spiritual purpose, that of divination. This is contrary to a simplistic evolutionary view of science as progression from unfounded beliefs to empirical ones. From the beginning objective observations were intertwined with metaphysical interpretations. Throughout the history of Chinese astronomy, we find great efforts made for accuracy in observation. Yet, from a modern scientific perspective, there could have been no relationship between the accuracy of the celestial observations and the accuracy of the resulting divinations. The Chinese thought there was, of course, as do contemporary Western astrologers who make use of the most accurate available data for their horoscopes.

The question of whether a divinatory system could possibly be verified scientifically is somewhat moot as there is no known mechanism to explain how it could work. Divinatory pronouncements are characteristically vague and thus cannot be evaluated for accuracy, even if one wanted to carry out such an experiment. This vagueness allowed errors to be excused as misinterpretations by the divining specialist rather than as failure of the method. With the *Yi Jing*, a further extenuating circumstance was the morality of the person being divined for; if he or she was not sincere, or was morally corrupt, favorable auguries would not apply.

## THE INTERRELATION OF ASTRONOMY AND ASTROLOGY

Despite the disdain in which astrology is now held by scientific astronomers, the two separated only gradually in recent centuries as the physical nature of the stars and planets became understood. Even before astronomy defined itself as a science, there were skeptics who derided divination by means of the stars and planets, finding it far-fetched to imagine celestial objects could influence life on earth. Once the planets and stars were recognized to be the same kind of entity as the earth and the forces determining planetary motion were understood, it simply became implausible to interpret their positions as predictive of human events. It was not that astrology was empirically refuted, but that the universe became less mysterious and scientific explanations supplanted magical ones.

Scientific astronomy has practical applications, for example for navigation, but an essential motivation for its pursuit has been curiosity. In ancient times, however, the primary motive for studying celestial motion was the assumption that it would enable prediction of human fate. Initially, only royal affairs were divined in this way but eventually less elaborate forms of astrology suitable for individuals developed (Ho 2003:153–164).

The notion that planetary positions can predict events and provide a key for analysis of human affairs, though vigorously rejected by modern astronomers, has proved a remarkably durable system from its origin in Chaldea (Babylonia) in the second millennium B.C.E. to its continuing use by millions. That astrology persists into our own scientific age is not evidence of its accuracy but does suggest that its tenacious hold on the human imagination is worthy of study. Astrology is a correlative system, one whose categories many find helpful in interpreting their experience. Contemporary astrology remains based on the same three-thousand-year-old system, though it has been considerably updated over the millennia, most recently by the superimposition of pop psychology upon the ancient lore. That astrology still serves as a means of categorizing people and events should not surprise us, given that astrology was originally developed for this purpose. The need to assess character and to find ways to cope has been present since the dawn of our species. Now the most popular matters for consulting astrologers seem to be romance and life changes such as moving or changing careers. Astrology is also a means of entertainment; many check newspaper horoscopes for this reason and do not take them seriously. The principle of "as above, so below" has perennial appeal.

Nearly all societies for which we have adequate records developed forms of astrology including Mesopotamia, Iran, India, ancient Greece and Rome, Ancient Egypt, and the Mayan culture of Mesoamerica. Given the universal human desire to anticipate the future, astrological diviners were often held in high esteem by the powerful and the "science" was diffused by itinerant practitioners traveling along trade routes. Because of the latter it is often difficult to know the extent to which astrological concepts originated independently. In China indigenous systems predominated over foreign imports but Indian astrology, due to the proximity of that country to Persian and Hellenic culture, was more eclectic. Chinese astrology is different from that of Middle Eastern origin in that the emphasis is on the stars near the pole, rather than those of the ecliptic.

Since the movements of the heavenly bodies were thought to represent the human realm, it is not surprising that the congruence of macrocosm and microcosm extended to the human body. In certain Daoist internal alchemy practices, the constellations were visualized as within the body. Western astrology, which seems to have originated in Mesopotamia, also has celestial/somatic correlations. Thus the head is ruled by Ares, the first sign of the astrological year, the neck by the second sign, Taurus; as the year advances, the correlations move down the body all the way down to the feet, ruled by Pisces. Each sign is also associated with emotional and character traits, even finger-nail shape (Hall 1995:15–32). Interestingly, after more than three thousand years, many of the ancient correlations persist in modern popular astrology, testifying to the powerful hold such ideas have on the human imagination. Astrology continues to appeal because it links the details of a person's life to the macrocosm and because it holds hope of assistance in difficult life-decisions. It describes a world in which spiritual factors continue to be present. Few want to live in a totally disenchanted world. Astronomy, though shorn of its concern with fate, continues to have spiritual content. Such phenomena as the big bang and black holes fascinate because of their symbolic power. We still feel that understanding the heavens has something important to tell us about the universe we inhabit.

Despite the prestige of diviners in traditional China, we should not image that people of the time were unable to recognize that many were rogues and charlatans. This is not unique to astrology; from antiquity to the present time, many have used the trappings of the science of the day to gull the unwary.

## Mathematical Astrology

In general, mathematical astrology in traditional China was reserved for matters of state. Though the systems utilized no longer seem plausible to us, divination records are of great interest because they give direct evidence of the concerns of daily life. Another use is in dating. Because Chinese historical records often include celestial occurrences such as conjunctions, comets, and eclipses, modern knowledge of celestial motion can be employed to date the astronomical event referred to in the early records (Pankenier in Mair, et al. 2005:24–26).

The need to interpret omens was the primary impetus for accurate observations of celestial objects. This, in turn, led to gradual discovery of the principles governing their motion. Ability to predict events such as eclipses, which advanced dramatically during the Qing, greatly reduced the risk that rebellious elements would use the events in propaganda against the emperor. Since the Qing emperors were not Han Chinese but Manchurian, have supplanted the Ming by conquest, they were despised by many of their Chinese subjects. Hence better astronomy was important for state security. It also benefited the public at large by making eclipses less frightening. From this we can see that what would now be regarded as a spiritual impulse—attributing human significance to the movements of astronomical bodies—was essential in the development of the science of astronomy.

While the elaborate systems of calculative astrology were applied only to matters of state, a simpler system known as fate-calculation or four-pillar astrology was widely used for guidance regarding private life. This based its predictions on the hour, day, month, and year of birth, each of which was identified with a zodiac animal and one of the five phases. Use of this system did not depend on accurate calculation of astronomical motion because it was not concerned with objectively observable events such as eclipses.

Mathematical astrology in China and elsewhere developed elaborate systems for calculating planetary positions but these sometimes diverged from actual observation because the calculations were an end in themselves; they were not usually checked against observation records.

Modern Western astrology is also based on calculation rather than observation. Most who do horoscopes are not able to recognize the constellations and planets in the actual night sky. However they have become sophisticated in using scientifically determined ephemerides. Divination however is judged by its answers more than by the process for determining them. Whether the calculations are correct is irrelevant, so long as those using it think they are. Hence astrological calculations only needed to yield results, and were not necessarily verified by observation.

## Observational Astronomy

For imperial use, accurate calculation was vital because the events in question, such as eclipses, were visible to all. The need to control interpretations of eclipses and the like was the reason for imperial support of astronomy but it seems reasonable to assume that, as now, scientists motivated by pure curiosity obtained government support by emphasizing practical utility. The metaphysical and political applications of astronomy did prevent observation from reaching a high degree of accuracy. We shall see how much was accomplished with the simple instruments available.

Beginning in the Yuan dynasty, Chinese astronomy was progressively influenced by that of the Arabs who had developed instrumentation to a high state. With science it is often difficult to determine how much was native to a country and how much came from outside because scientific knowledge and methodology is universal. The same scientific discovery can be made anywhere; discoveries are not culturally specific, even though the mode in which they are expressed may be. National pride leads countries to emphasize their specific contributions, understandably. However, science is probably never purely the creation of one culture; each makes use of the discoveries, major and minor, of others. Hence when discussing astronomy in China there is often uncertainty about the extent of foreign influence.

## THE GNOMON

The gnomon is undoubtedly the most ancient astronomical instrument. Consisting of no more than a tall stick placed in the ground, with its perpendicularity established by use of plumb lines, this homely device permits determination of the summer and winter solstices. In his discussion of the gnomon (SCC 3:284–310), Needham cites evidence from oracle bone inscriptions that it was used during the Shang dynasty at least as early as the fourteenth century B.C.E. There is also a tradition that an observation was made by the Duke of Zhou circa 1100 B.C.E. The earliest explicit textual reference is in the Zuozhuan commentary to the *Spring and Autumn*

*Annals*, which explains that it is important to be able to identify the solstice and equinox so that auguries of "clouds and vapors" could be noted (SCC 3:284).

At the time of the summer solstice, because the sun is then at its highest point in the sky, the pole's shadow at noon is the shortest of the year and, reciprocally, it is longest at the winter solstice. The longest day of summer seems to have been identified at times by using a template, eventually made of jade. The length of this template may have been used as a standard of measure of length but would be arbitrary since it is determined by the height of the gnomon itself.

Much more information can be obtained from this simple pole. Since the angle of the sun is dependent on how far the gnomon is from the equator, the relative terrestrial latitude of the pole can be determined, though this was only approximate since neither the size of the earth nor its curvature were known. That the gnomon was used for such purposes reminds us that there was interest in the earth and sun as physical entities with measurable qualities, not simply as subjects of myth.

Over Chinese history, use of the gnomon persisted but its design became more elaborate. Near Loyang, an ancient capital, there is a Ming restoration of a Yuan dynasty tower containing a forty-foot gnomon. A horizontal masonry structure north of it was used for measuring the length of the shadow (SCC 3:297). The larger size permits more accurate measurements since the changes of the height of the sun from day to day will result in larger changes in the length of the shadow. Records of gnomon observations have enabled modern historians to calculate the location of ancient sites. If the height of the gnomon and the length of the shadow at the solstice is recorded, latitude can be calculated.

Sundials were also used in China and in a sense are adaptations of the gnomon for measurement of the sun's daily, rather than yearly, movements. By tilting the angle of the upright beam so that it points toward the pole, the rate of movement of the shadow will be constant, allowing measurement of time of day. After the development of the compass, this latter instrument was used to properly align small portable sundials, thus increasing their accuracy.

Other methods of measuring time were used in traditional China including the water clock that was mentioned in *The Book of Rites* (*Zhou Li*), one of the five classics written in the Zhou dynasty (SCC 3:319). In Buddhist monasteries incense sticks in straight or spiral shapes that could measure much longer durations were used to time meditation sessions and other activities of the monks or nuns. Thus incense simultaneously served to honor the Buddha, or other sacred figure, and to organize human activities. Time has spiritual significance in all cultures but is not simply spiritual; it was equally used for scheduling military activities.

## THE SIGHTING TUBE AND THE ARMILLARY SPHERE

In this era of complex instrumentation, it is well to be reminded how much could be learned by human intelligence with the simplest instruments. The attention paid to the determination of the solstices shows that meticulous observation of cosmic events coincided with the beginnings of civilization more than three thousand years ago. The Chinese never developed optics and thus could only observe with the naked eye. Nor did they recognize the heliocentric geometry of the solar system. Yet they were careful observers and did attain an understanding of many of the fundamentals of the movements of celestial bodies. The sighting tube, no more than a narrow tube through which a star or planet could be observed and its angle with the observer's position on earth measured, extended the knowledge of heavenly movements considerably over what could be determined by the gnomon which was mainly suited for observing changes in the position of the sun. The sighting tube is mentioned in the *Huainanzi*, indicating that it was in use by the late Warring States or early Han dynasty. Needham suggests that the primary early use of the sighting tube was to locate the true north celestial pole. This is plausible as the pole had considerable religious significance in early China. As the fixed point about which everything else revolved, the pole star represented the emperor while nearby stars symbolized his high officers (Ho 2003:140). The big dipper, called the plow, was used in Daoist ritual and in feng shui determinations. An interesting ritual, still in use in Taiwan, involves an elaborate series of steps tracing the positions of the stars in the plough.

The armillary sphere is not a closed sphere like a modern star globe but rather a group of rings graduated with scales to record positions of celestial bodies. This invention is not unique to China but is illustrated in Ptolemy's *Almagest*. The Chinese armillary differed from those of Arabia and Europe in that its orientation was in relation to the equator rather than the ecliptic. Chinese astronomical texts as early as the Han describe specific famous armillary spheres in a way that suggests that only a few existed at any given time. It is clear however that some existed in private hands, despite recurrent imperial suspicion regarding astronomical knowledge outside its control.

## ASTRONOMY AND REGULATION OF HUMAN ACTIVITY

The development of armillary spheres indicates that the Chinese valued precise quantitative measurements. They were also aware of measurement discrepancies and put serious effort into reconciling inconsistent data. Yet the motivation seems more spiritual than scientific. We get a good sense of the concerns of Chinese astronomy from the *Huainanzi*, a text of the early Han compiled sometime before 139 B.C.E. (Le Blanc 1993:189). Le Blanc summarizes the work as having as its overriding concern the attempt to define the essential conditions for a perfect socio-political order. Examination of the Huainanzi chapters on astronomy and on seasonal rules confirms this characterization and gives us insight into the primary concerns of Chinese astronomy, namely providing indications of heaven's moral approval or disfavor. The "Treatise on the Patterns of Heaven" contains many highly specific observations, for example, "After fifteen more days, (the handle of the Dipper at midnight) points to jie:

This is the Awakening of Insects node. Its sound is like (the pitch pipe) Forest Bell. (Major 1993:88)

Two points are of interest here. First, there is careful observation of the seasonal positions of the Great Dipper in relation to the seasons. Since the position is at midnight, some accuracy of timekeeping is also inherent. However the correlative mode is prominent with each celestial position and season correlated with a specific musical note.

Another chapter, "The Treatise on Seasonal Rules" (Major 1993:217–268), represents a Chinese almanac giving correlations for each month and, most significantly, specifies proper activities for the ruler and his retinue for each time of year. Each of the twelve chapters follows a stereotyped format in providing twenty-four categories of correlations that include the sun's position, appropriate god, note on the pentatonic scale, location for palace animal sacrifices, which of the unfortunate animals' organs should be offered first, expected omens, appropriate room for the emperor to occupy, music to be played, color of clothing for court ladies and warnings of the dire consequences likely if activities proscribed for that month are carried out (Major 1993:220f). This represents the extreme of the tendency of the ancient Chinese to regard each time as having its own quality for which certain activities are proper and others to be avoided. The fixity of the system is unlike modern ways of humanizing time, which tend to be individual and spontaneous, such as "I'd like to go to a movie," or, "I'm in the mood for Chinese food." There are more serious implications of time, for example, having to be at class or at work. Moderns tend to dislike having their lives rigidly regulated by time, but if this was so in traditional China, there is little record of it. Perhaps there was little dissent because it was generally believed that to ignore these prescriptions and prohibitions was to risk grave misfortune, not only for the individual but also for the empire as a whole. While the *Yi Jing*, has a certain fluidity in that each casting of a hexagram can have any result and interpretation is conveniently open-ended, the seasonal rules are quite rigid. For court officials, disobedience to them would have been considered serious disrespect for the emperor, with possibly fatal results.

Bizarre as these rules now seem, we should understand that not just China but all societies have a deep anxiety concerning the possibility of disintegration of order. Three thousand years later we think of social order in quite different terms and try to maintain it by means of economic and social justice. In traditional China, order was maintain by following rules supposedly determined by heaven. In the religions of the Book, behavioral rules are attributed to a theistic God. Otherwise the kosher dietary rules, Sabbath, partial fasting during certain seasons, and others are not unlike the seasonal rules of China. The trend has been to liberalize these rules, though some choose to follow them voluntarily. No society can function without rules and all need a form of authority for their rules. For America this is the consent of the governed, at least in principle. In traditional China the rules were as much a part of the natural order as sunrise and sunset; the emperor's authority was part of this natural order. In the romantic and New Age there is a new desire to live in accord with the natural order, rules, but this is conceived quite differently than it was in the *Huainanzi* and other Chinese sources.

## ASTERISMS AND CELESTIAL MYTHOLOGY

It is inherent in the human brain to organize perceptions into coherent patterns; indeed it is impossible for us to avoid doing so. Hence all cultures have found patterns in the stars onto which they projected their myths. As in ancient Babylonia, so in early China, stars were perceived in terms of asterisms or constellations, which conferred order on the otherwise chaotic near-infinity of luminous dots in the night sky. Inevitably, some stars seem to fall into discreet groups. Most conspicuous of these is Ursus Major, the big dipper, referred to as the plow in China. This was the most important asterism for the Chinese who tended to have a polar orientation in contrast to the ecliptic orientation of early astronomy in Babylonia, Egypt, and Greece. From these regions eastward to India, there is a quite surprising degree of continuity across time and place. In contrast, Chinese astrology developed independently until Western forms entered China with tantric Buddhist monks in the third century C.E. (Ho 2003:69). While it is an oversimplification to refer to all astronomy from Greece to India globally as "Western," the various forms in this wide geographic expanse, regional forms do share key features such as the constellations of the zodiac, twelve houses and similar, though not identical, mythological associations with the planets. Thus Mars represents war and is generally malefic, Venus represents love (though once war as well), and pleasure, and so on.

When I studied astronomy in my early years, books took pains to explain that the mythological correlations of the constellations were specious. This dismissal was due to failure to appreciate early ways of thought. While it is obvious that mythical associations of celestial bodies have no meaning in scientific astronomy, they are of great interest in the objective study of early cultures. More nuanced concepts of mythology see it not as collections of fictions to be refuted but as evidence for how earlier humans accounted for matters of great concern such as the origins of the universe, human beings and civilization, as well as natural phenomena such as the markings on the moon. Myth can be studied scientifically, even though not itself scientific. The work of Carl Jung, though not rigorously scientific in method, has been highly influential in presenting myth as a route to recognizing deep patterns inherent in the consciousness of the human species.

## THE JOURNEY TO NOT EVEN ANYTHING

While the seasonal rules suggest a spirituality concerned with regulating human behavior, other strains of cosmological thought used the vast celestial realms as a metaphor for the freedom of human imagination. Particularly prominent in traditional China from the Warring States onward is the theme of the journey to other realms. Here are two examples from the eccentric Daoist philosopher, Zhuangzi:

Why don't you try wandering with me to the Palace of Not-Even-Anything... our discussions... will never come to an end, never reach exhaustion.
. . .
I ramble and relax in unbordered vastness; Great Knowledge enters, and I don't know where it will ever end. (Watson 1968:241)

These quotations from one of the two great Daoist philosophers express a spirit as different as can be imagined from the seasonal rules of the *Huainanzi*. Rather than advocating restriction and minutely ordered behavior, Zhuangzi appeals to the imagination and its ability to wander anywhere. Nature is seen as open-ended, so vast it can never be fully comprehended. This outlook is closer to that of science than that of the *Huainanzi* in its recognition that knowledge is never completely settled or complete. While Zhuangzi emphasizes nonverbal or intuitive knowledge, he advocates being open to all things. While he does not directly challenge the standard metaphysics of yin and yang, he does not subscribe to the sort of rigid cosmology that became orthodox in the Han and subsequent dynasties.

Some modern exegetes have worried that the *Zhuangzi* (that is, the book attributed to this philosopher) was anti-intellectual and disparaged knowledge. An example of what can be construed in this way is:

Not to understand is profound; to understand is shallow. (Watson 1968:243)

Certainly if taken literally, this would seem to advocate ignorance. However, it is more likely that what Zhuangzi is admonishing is resisting accepting any form of knowledge as final, that admitting what one does not know is a more complete form of knowledge than deluding oneself that one's understanding is complete. Here is a similar statement from the great contemporary physicist Freeman Dyson:

As a scientist, I live in a universe of overwhelming size and mystery.... Behind the mysteries we can name, there are deeper mysteries that we have not even begun to explore. (Dyson: 2002)

The *Zhuangzi* shows that the spirit of open inquiry was part of Chinese culture from quite early. This is also apparent in the inventiveness of both traditional and contemporary China. The idea of "Flights Beyond the World," that is, wandering in imaginary realms, remained prominent in later Daoism (Schafer 2005:234–269). Here is an example, from the Tang poet Po Chu-I:

He pushes the clouds aside, harnesses the winds, courses like the lightning. Mounts the sky, enters the earth, looks for [the legendary beauty Yuan Guifei] everywhere.
　　Exhausting the blue gulf above and the yellow springs below.
　　(Schafer 2005:246; translation slightly modified)

One sign of the greater degree of mental freedom implied in these works is that female travelers and deities are more common in these journeys through the void than are female figures in Confucian and Buddhist writings. Thus Li Bai, one of China's greatest poets wrote of a friend who was a Daoist priestess:

　　　　Under her feet far-wandering shoes
　　　　To skim the waves, raising pure-white dust
　　　　To find the transcendents heading for the southern sacred
　　　　　　mountain
　　　　Where surely she will see the Dame of Wei.
　　　　(Schafer 2005:237; translation slightly modified)

Daoist rituals also made extensive use of star imagery. "Star hats" are frequently mentioned in descriptions of rituals, though none have survived (Schafer 2005:225). Many paintings depict Daoist deities of stars, planets, and the sun. Examples of Daoist visual representations of cosmological deities have only recently attracted the interest of Western art historians. Interesting examples include the Emperor of the Pole Star, the Lords of the Root Destiny Stars, and the Gods of the Twenty-eight Lunar Mansions (Little and Eichman 2000, Plates 76, 78, and 79). Daoist celestial deities also appear prominently in the Buddhist genre of sea and earth paintings, intended to represent all creatures of the cosmos; belief in these supernatural beings was not limited to Daoism. Although the deities depicted are standing on clouds, there is otherwise nothing astronomical in these paintings. Simple representation of asterisms consisting of circles connected by lines, very much like modern diagrams of constellations, are frequent in magical diagrams and in some of the graduated circles of the *luopan*, the compass used for feng shui (geomancy). Buddhism also developed an elaborate cosmology but rather giving stars and planets a central role in its mythology, tended to create near-infinite Buddha realms with little reference to actual visible celestial bodies. It is often remarked that Daoism is concerned with the natural world while Buddhism disparages it. This is at best a partial truth, because Daoist imagery, though it often starts with natural phenomena, ends up with abstract metaphysical diagrams on the one hand and anthropomorphic deities on the other, both far from what can be observed in the natural world. The stars were also part of Daoist "internal alchemy" or longevity practices, which included methods of pulling the energies of stars into the practitioner's body.

This use of celestial imagery is, of course, not science but is the earliest form of speculation about what might lie beyond the visible. As we have often noted previously, fitting the observations into the preexisting metaphysical cosmology deflects attention from observing them as part of objective nature.

## CONFUCIANISM AND CHINESE ASTROLOGY

In Daoism, celestial bodies represent escape from the limitations of the world, whether in the literary creations of Zhuangzi and the poet Li Bai, or the spiritual exercises intended to surmount the greatest of human limitations: mortality. This is in keeping with the this-worldly Daoism of Laozi and much of Zhuangzi that teach avoidance of social limits—though not the do-you-own-thing sort of rebellion that flowered in the sixties. In contrast to the Daoist idea of space as a place of carefree exploration, Chinese official astrology had, as would be expected, a clear Confucian emphasis on correct behavior. Shafer has assembled examples of astrological interpretations of comets and other celestial events from the Tang. Here is his summary of the response to the appearance of Halley's comet in March of 837 CE:

On the Advice of the astronomer Royal Chu Tzu-jung, the daily allotment of food in the palace was reduced to one-tenth of normal, and all construction on the palace grounds was halted "in response to heaven's denunciation." . . . On 16 April, classical musicians were dismissed from the palace, and on the following day the young monarch issued a decree outlining the grave implications of the dreadful apparition, and announcing the need for further austerities, soul-searching, and penance. (Schafer 2005:113)

The interpretation of comets as unfavorable omens was by no means unique to China. As often with prognostications, ill consequences of improper behavior are to be averted by austerities (Terzani 1997). Here again we note the anxiety regarding maintenance of order that is prominent in Chinese writings from the oracle bones to the Little Red Book of quotations from Chairman Mao.

The imperial government's approach to counteracting the destabilizing effect of celestial anomalies was never to deny that they indicated heaven's judgment of the administration. Rather, it was by ingenious interpretation so as to present the anomalies are favorable, and especially in the late Ming and Qing, by advances in astronomy that made possible more accurate prediction of such phenomena as eclipses. As astronomers were government employees, no one outside official circles would know of such events in advance. Thus the imperial authority maintained control of the interpretation of eclipses and comets so that they would be presented as auspicious signs favorable to the emperor. High priority was given to supporting research leading to more accurate calculation of these events.

Always the response to comets and similar phenomena was a call for austerity on the part of the emperor. At least one text mentions reduction of penalties for those convicted of crimes so there is a sense that mercy might be an appropriate way to regain the mandate of heaven. Other than this however, it was the emperor's personal and ritual behavior rather than his policies that was assumed to be the cause of heavenly disfavor.

It is easy to assume from textual records, such as that quoted above, that astrological advice was accepted without skepticism. As noted previously, this is probably incorrect. It seems probable that imperial advisors interpreted omens so as to advance their own interests, or simply to avoid offending the emperor, who held immediate power of life or death. A few textual sources express skepticism as to the role of the stars in earthly affairs or the honesty of interpretation by court astrologers. Nonetheless, the main impetus for study of the celestial bodies throughout Chinese history was their assumed connection to human affairs.

## THE CALENDAR

Calendars seem but banal necessities today, though all of us are quite dependant on them for the temporal organization of our lives. Since the Gregorian calendar to which we are accustomed, functions without any apparent glitches—other than the mild inconvenience of leap years—it is necessary to be reminded that it took millennia for accurate calendars to be devised. The great problems of calendar making are the fact that the year cannot be divided into days without a remainder and the incommensurability of the solar and lunar cycles. The seemingly simple innovation of adding a day every four years prevents the drift of months that disrupts their correlation with the seasons. That the lunar and solar years are not completely in phase is no longer noticed. The lunar year is simply a curiosity only thought of when needed for determining traditional Asian festival days, notably the Chinese New Year, now celebrated worldwide.

The basic problem with calendars was not the difficulty in determining the exact length of the solar year, but the numerological unattractiveness of fractions. Accurate calendar development was inhibited by the unpalatable truth that the cycles of solar days, lunar months, and solar days are slightly out of phase. Traditional thinking favored numbers with many factors, because this seemed to have mystical significance. Thus we often find calendrical representations that reduced the number of days in the year from 365.2424 to 360. This is Babylonian in origin and persists in the mathematical convention of dividing circles into 360 degrees, an appealing number because it has many factors.

The three major astronomical cycles of the earth's rotation (day and night) the moon's phases and the rotation of the earth around the sun cannot be divided into each other without remainder. Of these, the rotation of the earth around the sun—and here we must remember that the heliocentric geometry of the solar system was not recognized—though a constant, was the most difficult to measure accurately. Moon phases are also constant in duration but days and nights are not. Mathematical efforts to divide the year by an integral number of days or lunar cycles could not be successful. A result that we find in the Chinese calendar is use of more than one system simultaneously. Ho (2003:30) notes that about one hundred different calendrical systems were used in China at different places and times. Only a few were widely used and since most lived their lives by the sun and the seasons, exact dates were not necessarily important. There are still many living Chinese who do not know their actual date of birth, for example. (Some make use of this to subtract a few years from their real age.)

Though most lived their lives without reference to dates, calendars had a place in agrarian society. On a social level, religious festivals needed to be held at times of year based on the event they commemorated. Since most of the populace was illiterate, holding religious events at particular points during the year helped inculcate the basic ideas of their religion. Thus calendars were essential means of political and religious indoctrination. Even now, certain days become something more than arbitrary numbers because of their associations with prior events of cultural importance, each with its own affective tone. This is quite familiar in Christianity, for example, in which Christmas is joyful, Easter solemn, and so on. Chinese festivals with spiritual significance include not only the New Year, but the midautumn moon festival and the hungry ghost festival (of Buddhist origin). Dates for these are set by the lunar calendar as in ancient times, when the moon's cycle was visible to all.

Since days were regarded as favorable or unfavorable for particular activities, a major function of calendars was to permit astrological determination of what days were proper to particular activities of the emperor, his officials, and others who did not simply till the fields. Over time, use of almanacs and other methods to select propitious dates for important life events became virtually universal. (When I was married in Hong Kong, a friend consulted an almanac to pick an auspicious day for us. Since the same almanac is widely used, we found that the most favorable day that month was fully booked in the wedding registry, while the inauspicious days were mostly vacant. Fortunately, my fiancée was able to use her connections to get a slot on a very favorable date. The prudence of this maneuver is suggested by the fact that we are still happily married more than sixteen years later.)

## THE SEXAGENARY CYCLE

The most pervasive system for expressing dates in China, which is still in limited use, is termed *ganzhi* in Chinese and generally referred to in sinological works as the stem-branch or sexagenary system. This used two elements for counting days, a series of ten *shigan*, usually translated as heavenly stems and another series of twelve *shi 'erzhi*, or earthly stems. The stems were employed from the Shang dynasty onward and represented a ten-day interval approximately comparable to our week. The stems (more accurately trunks) and branches are by analogy to trees; thus time was represented by a spatial analogy as indeed it always has been, until the recent development of digital displays and atomic clocks. Time cannot be measured directly and so was determined by spatial change: the movement of the sun, water in a clepsydra, burning of incense, or the hands of a watch (Wilkinson 1998:201).

The week, whether of seven or ten days, seems to be a usual interval for short-term planning. Oracle bone divination was regularly carried out to determine whether the ensuing ten-day week would be favorable (Wilkinson 1998:179). When paired sequentially with the branches, a series of sixty pairs was generated. From the time of the Shang dynasty oracle bones onward, many documents were dated with this sixty-day cycle. Years were also counted in this way, generating a sixty-year cycle. Obviously each stem–branch year designation would recur every sixty years. However, the sexagenary date was combined with the reign name of the emperor so that ambiguity was avoided. Dates based on the reigning king are common in traditional cultures and are another means of connecting abstract time with human events. Other systems were used at various times and places in China, including one based on the twelve annual stations of the planet Jupiter and even on a fictitious counterorbital analogue of Jupiter (Wilkinson 1998:183).

The twelve-year animal "zodiac," was a somewhat later development being apparently introduced during the Qin dynasty. In this system, most familiar from Chinese restaurant placemats, each animal designates an entire year. It is not analogous to the twelve-month Western zodiac because the animals do not designate positions of the sun on the ecliptic. The animal zodiac also generates a sixty-year cycle by associating each animal in a given year with one of the five phases (*wu xing*). Thus in each sixty-year cycle, there will be five years of the horse, each associated with one of the five phases: earth horse, metal horse, fire horse, etc. Within this, days are specified by stems and branches, and hours (equal to 120 minutes) by the same twelve animal sequence. This is the basis of such poetically named times of day as "The hour of the rat" (11 P.M. to 1 A.M.). Given that these cycles are in phase only every sixty intervals, calculating the correct stem, branch, wu xing phase, and animal would be laborious and difficult. Fortunately tables, entitled "Ten Thousand Year Books" have been readily available at least since the Ming. (The title actually means "Many Year Book," since the character for ten thousand also means "many.") For those curious about the details of this elaborate system, an English version of the tables is available (Sung 1999).

That each time had a name rather than a number is consistent with the Chinese view of time as qualitative. Wilkinson (1998:172) comments as follows:

Chinese traditional society was above all one in which the highest importance was attached to choosing the right time and the right place for activities both high and low, solemn and trivial. The selection was done using a myriad mantic arts.

We have already seen how time determined in minute detail what social behavior was appropriate. Because everything was interconnected, to deviate from any detail was to threaten the entire social fabric. Even using the wrong sauce for a particular dish was a threat to the natural order, though one imagines that such *gaffes* were commonly overlooked. The comparatively amorphic quality of modern life in which, for example, meals are eaten when one happens to have a free moment, gives rise to nostalgia for the more structured time of traditional societies. We should not forget, however, that the rigidities of correlative thought often complicated what for us are simple activities.

The Chinese concern with auspicious dates is a response to the unpredictability of life. We all know that actions that may be successful at one time may fail at another. Divination to reduce this uncertainty seems to have been universal in traditional cultures, though not without its skeptics. Though reduction of uncertainty is a major motivator for science, there is no continuity between divination and empirical science. Scientific weather forecasting or MRI has no relation to oracle bone divination or Chinese pulse diagnosis, to give but two examples. Divination and science to some degree respond to the same human needs but in incompatible ways.

## DIVINATION DOWN TO EARTH: FENG SHUI

Proper selection of grave sites has been a preoccupation of the Chinese from the time of the earliest archeological records. Earth-based feng shui eventually mostly based astronomy as the method of determining favorable siting, though virtual stars were used in calculations. The motivation for grave-siting has always been a concern to placate the spirits of the dead lest they take revenge by adversely affecting the affairs of the living. This constituted the primary use of the *luopan*, a compass with multiple concentric dials representing the same metaphysical elements we have been discussing: yin-yang, wu xing, stems, branches, constellations, *Yi Jing* trigrams, and magical Chinese characters. It is easy to see why the compass would be used for such a purpose as the movements of the needle must have seemed quite mysterious and do in fact respond to an unseen force. Like its related art of calculative astrology, feng shui as practiced now does not involve direct observation of stars, some versions make use of what can be termed "virtual stars," such as the "flying star" system which is also based on the *Luoshu* magic square. As an example of the complexity of such systems, one modern textbook of feng shui provides an appendix of over six hundred flying star charts for the years 1864 to 2043 (Koh 2003).

While the system of one form of feng shui makes use of the compass for orientation, it is not empirical in any other way. The elaborate calculations are not based on any actual empirical observation of the effects of dwelling place on human welfare. Some modern principles, such as not living in the face of oncoming traffic, such as with a T-shaped intersection, follow common sense but such instances are limited. Despite this lack of evidence of efficacy, feng shui, like Chinese food, has spread all over the world. One can speculate on the reasons for this. First, just as the *Yi Jing* humanizes time, so feng shui humanizes space. In our homes, each of us has favorite places to sit and other areas we tend to avoid. Pets do the same. Though the subtle emotional reasons for these responses are difficult to pinpoint, they do mean that how a living space is arranged affects our well-being. Feng shui offers explanations for these feelings and also helps at the difficult moment of facing an empty room and trying to decide how to arrange its furnishings. If a system helps with the attainment of comfort, it serves a useful human purpose, however unscientific its basis. Science is not much help in interior decorating, though it is essential in safety considerations such as avoidance of asbestos insulation. Feng shui, like any other form of divination is not innocuous if allowed to override common sense. Thus some Chinese spend immense sums to counteract bad feng shui as a cause of ill health or business troubles when the money and effort might better be spent on medical care or management advice. Yet feng shui continues to have a powerful hold on the Chinese imagination. For some, the ratio-nale is like Pascal's wager, which proposed that since God might exist it was prudent to conduct oneself as if He does. The Chinese version is that feng shui probably is not true, but just in case it is, why not follow it? The assumption that modern science will drive out "superstitious" beliefs like feng shui has turned out, so far at least, to be wishful thinking. One reason is that feng shui is both the cause and the means to relief of anxiety about locating a home or business, or grave. This is usual in religion, most no-tably Christianity, Chinese folk religion, and some forms of Buddhism that invent hells, then offer the only means to prevent oneself from ultimately falling into them. The question of whether the net effect of religion is to increase or decrease human fears may not be answerable, but one cannot help but notice that many choose religion of their own free will. It is not only childhood indoctrination that makes people religious.

Obviously proper location of buildings is important but the extensive lit-erature on feng shui does not suggest efficacy in avoiding sites that would be subject to natural disasters as floods and earthquakes. The concern was less natural hazards than supernatural ones.

## THINKING ABOUT CHINESE ASTROLOGY

We have seen that until relatively late, Chinese astrology was primarily employed as a means of aiding decisions of state, especially with regard to war, and as a means of political manipulation by cunning interpretation of celestial portents. Before dismissing this as mere superstition, the nature of human knowledge in prescientific times should be considered sympathetically. First, the rules coordinating court activity with astronomical patterns, while hardly democratic, does show that even emperors could not do exactly as they wished but were expected to conform to a moral pattern that was not arbitrary but was inherent in the nature of things. (This, of course, did not mean that all emperors were moral, nor that this ideology could not be twisted to claim that improper behavior was actually sanctioned by heaven.) Second, because imperial astrology was based on actual positions of celestial bodies, it was in a sense the most advanced science of its time. Certainly Chinese astronomy was far more empirical than was the medicine.

Government support of the costs of the astronomical bureau for guidance on important decisions indicates the seriousness with which matters of state were deliberated. Our methodology today is far more scientific, but far from infallible. Despite satellite photos and the like, intelligence about the military capacity and intentions of other countries has often turned out to be disastrously wrong. Modern government policy depends in part on econometrics which, like astrology, is dependent on number-crunching. While policy that utilizes these statistics has greatly reduced economic misery, it has hardly abolished it. Our most sophisticated models cannot forecast such parameters as currency and commodity price fluctuations with sufficient accuracy. This is not to equate economics with astrology but simply to emphasize that the governments of traditional China, like governments today, utilized what they thought were the best methods available and that they were based on science, though the mode of interpretation we now know to be fallacious.

I do not mean to take this point too far, however. Government policy used what in science are referred to as substitute endpoints. If people were starving, rather than focus on what interrupted the supply of food, the cause was presumed to be extravagance in court displays or excessive sexual activity on the part of the emperor. (How excess is to be defined for men who typically had more than one hundred wives is unclear.) There were practical efforts to improve control of water resources and methods of agriculture. However, the assumption that personal propriety affected natural events greatly inhibited objective measures to benefit the population. While such corrective measures as the emperor reducing the number of court musicians seems absurd now, it was never seriously questioned.

In China, important bits of accurate scientific knowledge tended to languish. Many observations of startling scientific insight ended up as metaphysics. The Chinese were well aware of the difference between regular (fixed) stars and irregular ones (planets). The great Song philosopher Zhuxi noted that the light of stars was direct, while that of the planets was reflected from the sun but went on to explain this, not by inference about the possible physical nature of the planets, but in terms of qi and yin-yang: "Regular stars . . . are what are [formed through] congelation (sic) of extra yang qi." (Kim 2000:143) This is a salient example of how correlative cosmology tends to close further observation by providing easy, but ultimately empty, metaphysical explanations.

Thus, astronomical observation did not lead even to hypotheses regarding the physical nature of stars and planets; the compass, though used for navigation, was mostly a way of determining qi for feng shui; and the success of vaccination did not lead toward concepts of infection and immunity. This should not be regarded as a shortcoming of Chinese civilization because such advances took a long time to be developed in the West also. It simply shows how difficult it has been for humanity to grow beyond correlative cosmology to empirical science.

## ASTRONOMY AND ASTROLOGY IN INDIA

As already pointed out, for everyone in the ancient world, the night sky and its cyclic changes must have been as familiar as they are now for amateur and professional astronomers. Indeed this experience must have been one factor in universalizing human thought; despite differences in geography, the patterns in the sky are nearly the same everywhere. Most celestial events follow regular cycles but a few such as comets and eclipses seem not to. It is human nature to expect things to be constant, even though we all know they are not. Thus we feel that when things change, an explanation is to be sought. The movements of celestial objects display the paradox of permanence and change; most of the changes in the sky follow expected patterns but some do not, just as is the case in human life generally. It is tempting to regard the experience of the patterns of the night sky, along with seasonal changes, as the inspiration for early metaphysical speculation, though this is impossible to prove.

To the ancients it seemed self-evident that changes in the sky were related to changes in human life on both the personal and state levels. We should not reject as foolishness the astrological view that changes in the heavens control human destiny because the early world was dependant on marginal agricultural production, which in turn was determined by the seasons. Changes in the movements of the sun were the most important single factor controlling human activity and comfort. For this reason great effort was made to understand and predict the events in the sky. In India as well as China, meticulous observations of great accuracy were carried out. Their study of astronomy yielded only information about the movements of the celestial bodies—because this was all that could be learned by direct observation. Geometric models in India, influenced by astronomy in Mesopotamia, Greece, and Islam were more accurate than those of China though the nature of the observed objects really only began to be understood with the invention of the telescope in Europe.

It was always assumed that objects visible in the sky had spiritual significance, not only for the reasons already mentioned but also because the sky is distant and inherently mysterious. Human imagination has always placed alternative worlds in the sky; this is as true today with science fiction as it was three thousand years ago. Heavenly bodies have had mythological associations for as far back as we have records.

The mythological interpretation of celestial objects did not inhibit observation; this was carried out in India, just as it was in China. That the interpretations tended to be astrological and mythological should not diminish our admiration for their work. While one must assume that simple curiosity was a motive for observation, it is clear that the primary motive was for accurate astrological predictions. Much of the work was government supported and then, as now, governments are interested in practical results, particularly those that can strengthen their grasp on power.

## CALENDARS IN INDIA

The problems of calendar development are the same everywhere. Defying the assumption that natural patterns fit those of the human mind, the solar year is approximately 365 days, an odd number with few integral factors, rather than the numerologically more appealing value of 360. In fact the number of days in a year is not an integer. Further complicating matters, the visually conspicuous lunar cycles do not divide evenly into the solar year. Thus calendar makers have to accept a compromise between numerological theories and actual observation. The closer the calendar is to metaphysically derived numbers, the greater the drift in dates over the years. In responding to this dilemma, Indian calendrical development went through three stages. The first, which has yet to be fully reconstructed, was based on the Vedic texts, particularly the *Vedanga Jyotisa*. As was the case in China, records of astronomical matters are mentioned in very early texts demonstrating the close association between spiritual concerns and patterns in the sky (Chakravarty 1997:168–171). The second phase was introduced during the Greek conquest by Alexander's armies, and the third was a return toward indigenous forms, though made more accurate by use of the observations by Indian astronomers. With some modifications, this latter calendar is still used today in India for festivals and for astrological purposes. The need for religious festivals and sacrifices to be made under specific celestial configurations was another religious motive for careful observation. Orthodox Hindus still use an almanac based on this calendar to determine appropriate observances for specific days.

This Siddhantic calendar used the Western zodiac, of Babylonian origin, and was based on the movement of the sun along the ecliptic. Indian astronomers also attempted to start the calendar at something like the Chinese "perfect epoch," a moment when all the planets were in conjunction at the zero point of the zodiac, that is, at the same longitude on the celestial sphere. This zero time was the start of the present epoch, termed the kaliyuga, extrapolated to have begun 4,320,000 sidereal years ago (Chakravarty 1997:170). Once again we find the Hindu fondness for very large numbers interfering with empirical deduction, just as was criticized by Al-Biruni. Chinese estimates of the "superior epoch" varied from 143,127 to 96,961,740 years ago. Eventually the effort to determine this was abandoned in China (Ho 2003:32). The motivation for this search for the origin of the kaliyuga was religious: the desire to find complete regularity in natural phenomena and the universal human wish to discover origins. While this encouraged great efforts in observation, the desire to fit the observations to religious expectations ended up interfering with lucid interpretation.

India developed astronomical instruments as did other ancient civilizations in order to increase the accuracy of observations. These included the gnomon, the armillary sphere, and others. It is unlikely that these developed entirely independently in different areas but the influences are now difficult to trace. In India, as in China, astronomy was closely associated with mathematics; trigonometry and geometrical diagrams were employed to understand planetary motions. Because of the need to coordinate religious events with celestial ones, immensely difficult observations were made with instruments that reached the limit of sophistication possible without knowledge of optics and electronics. Elaborate calculations with correction factors were made to try to improve the accuracy of predictions. However, Indian astronomers seem to have been content with observation and calculations that would predict celestial events and did not advance in understanding the structure of the solar system, which was still assumed to be geocentric. We can see that astronomy was stimulated by a variety of human concerns: to help anticipate the future, and to better understand the universe both metaphysically and empirically.

A notable astronomer was Brahmagupta (b. 598) who developed a method for calculating the motion of a planet and even a means to calculate parallax, the distortions in measurements introduced by the method of observation. The latter shows true empiricism since any correction of calculations requires comparison of the results actually obtained with those predicted by theory (Sarma 1997:116).

## ASTROLOGY IN INDIA

*Jyotisa*, Vedic astrology, continues to be used in India and has attained a limited following in the modern West. Even today in India, comparison of horoscopes is an essential step in determining mutual suitability of prospective marriage partners. Given that marriages have been, and usually still are, decided by the parents, compatibility is hard to judge on a personal level. We have no data on whether selection of marriage partners by astrology is more successful than personal preference.

Like other forms of astrology, that of India employs complex calculations. It is clearly related to Western astrology, which originated in Babylonia and has been conserved to a remarkable degree for more than two thousand years (Holden 1996:5). The 360-degree circle, the twelve-segment zodiac, the house system in which the effects of planets depend on their location at the exact time of birth are all fundamentally similar. While this system did reach China from India and Central Asia via the silk road at least by the eight century C.E., the Western zodiacal astrology never supplanted the mostly polar-based Chinese systems.

While the system that originated in Babylonia has become increasingly complex and the nature of interpretations has been affected by cultural context, it is striking how little the basic elements have changed. Since there is no basis for assuming that astrology is correct, it provides a striking example of a set of ideas having great survival value, despite lack of objective verification. This is not to dismiss astrology as merely ignorance. Some astrologers were, and are, quite learned and it is possible that those consulting them benefit by having diverse possibilities suggested to them.

# Chapter 7

## Ecology and Environmental Consciousness in China

We must begin by correcting a common misapprehension about Chinese and Indian behavior with respect to the environment. Here are examples of this fallacy:

First, in regard to China:

[Daoism] believed that if left to itself, the universe proceeds smoothly according to its nature...Mankind's efforts to change or improve nature only destroy these harmonies and produce chaos.

   Simply put, feng shui is an early form of ecology, conservation, and environmental science. It unites heaven, earth, and humans in such a way that a respect for nature is developed along with the ideas of renewability and sustainability of resources. (Berger 293)

With respect to India:

The philosophical and scientific ideas developed in India over the centuries are, on analysis, found to be deeply related to ecological issues...(Chattopadhiyaya 1997:295)

   Such claims, appealing as they are, give a quite mistaken impression of the state of environmental awareness in both cultures. To begin with, there is no reason to think that Chinese or Indians have been any more careful of the environment than members of any other culture. Certainly a visit to a city in either country today quickly disabuses one of such a notion. Conditions are comparable to those in Europe or North America during the heyday of heavy industry. Both quoted passages contain a fallacy all too common in writings about Asian philosophy—mistaking the ideals of a culture for its actual behavior. The more remote in time or distance a country is, the easier it is to believe that they have avoided the problems of the modern world.

The first statement that Daoists recommended not going against nature is quite true. However this does not mean that economic development has proceeded in accord with these commendable principles. The second statement, about feng shui, is simply false. The purpose of feng shui was, and is, choosing sites and arrangements conducive to good fortune, getting rich and living as long as possible. This has little to do with the idea of protecting a fragile environment. Indeed, Chinese sometimes incur resentment in suburbs in Canada and the United States by cutting down attractive trees in front of their houses because they were told by a feng shui master that the position of the tree would block money from coming in. With India, we have the same fallacy of mistaking the ideals in religious texts for actual behavior. This inconsistency between ethical guidelines and actual behavior is hardly unique to Asia; it is universal in human societies. For this reason, it is important to be clear about whether actual or ideal behavior is being considered.

Recurrent in our examination of the relation of humans to their environment in traditional China thought will be paradox. As expressed vividly by the great scholar of Chinese ecology, Mark Elvin:

A paradox thus lay at the heart of Chinese attitudes toward the landscape. On the one hand it was seen, not as an image or reflection of some transcendent being, but as a part of the supreme numinous power itself. Wisdom required that one put oneself into its rhythms and be conscious of one's inability to reshape it. On the other hand, the landscape was itself tamed, transformed, and exploited to a degree that had few parallels in the premodern world. (Elvin 2004:322)

Elvin goes on to make a vital point, often ignored by those studying the writings of the past, and particularly by those attempting to find in them parallels of modern spiritual views:

This paradox shows that the relationship of a representation to a reality ... can at times be the opposite, in a sense, of what is. Behind all such studies of the perception of the environment there lurks a trap: that of assuming that people's actual *behavior* was on the whole necessarily in accord with the *ideas and feelings* expressed in our sources. (Ibid. 322)

This is obvious enough with present-day events because we have opportunity to compare fine sentiments with actual behavior. Many historical sources, especially spiritual writings, tend to present the ideal with little record of the actuality. There are of course writings from a disillusioned point of view. Thus Laozi, the author of the *Dao De Jing*, according to a famous legend left the civilized world disgusted by humanity's failure to accord with the Way. Yet even if we imagine that this appealing account is true, the story tells us about Laozi's idealism, not that of the millions who later claimed to follow his teachings. Similarly, Confucius as depicted in the Lunyu was constantly criticizing the propriety of others but this tells us nothing of the extent to which his admonitions favoring sympathy and mildness were actually followed by ostensibly Confucian officials.

As we take up the state of ecological awareness in traditional China, we must make a clear distinction between two contemporary usages of the term "ecology." In its initial usage the term referred simply to the scientific study of the interaction of organisms with their environment. Later, with the beginning of the modern environmental movement inspired by Rachel Carson's *Silent Spring*, the term has been applied to the social movement advocating protection of the environment against human desecration. Ecology in this sense goes beyond the purely scientific component to a set of ethical principles regarding the natural world as well as political activism promoting this ethic. While the second use of the term implies concern with scientific evidence for environmental harm, the first simply refers to objective study unrelated to any specific agenda. Unbiased interpretation of research on ecology is unfortunately uncommon, given that most of those involved have economic or ideological agendas.

When we consider environmental concerns in traditional China, we find the same dissociation between spiritual ideas and practical that we have been discussing. The early Daoist philosophical texts, the *Dao De Jing* and the *Zhuangzi*, contain much that has been a source of inspiration for the modern environmental movement. Yet one is hard put to find evidence that these texts, enormously influential as they were, played any role in fostering environmental consciousness in the modern sense. Rather, efforts at water and soil management stemmed from a sense of responsibility on the part of officials for the well-being of the people they oversaw. Thus to the extent there was a Chinese environmentalism, it stemmed from the practical side of Confucianism, not from the more abstract principles of Laozi and Zhuangzi. The concern of officials was for human well-being, not for nature itself. The modern environmental movement is also concerned with keeping the earth suitable for human habitation but extends it concerns farther, for example in trying to protect endangered species.

The two Daoist philosophers, particularly Zhuangzi, counseled withdrawal from involvement in government to a supposedly simple rural life, but this was for one's own protection against political misadventure, not for protection of the environment. Here are some examples of passages from the *Dao De Jing* that are adduced to support a theory of ancient Chinese ecological awareness:

> Love the world as your own self; then you can be trusted to
> care for all things. (Chapter Thirteen)

> Man follows the earth,
> Earth follows heaven,
> Heaven follows the Tao.
> Tao follows what is natural (Chapter Twenty-five)

> The universe is sacred,
> You cannot improve it.
> If you try to change it, you will ruin it. (Chapter Twenty-nine)

> The world is ruled by letting things take their course.
> It cannot be ruled by interfering. (Chapter Forty-eight)

It is easy to imagine that passages like these contain something resembling modern ecological consciousness, the view that the earth, and human society, function best when not meddled with. However, while the social order was seen as fragile, this was less clearly true of nature. Rather, while the natural order is the proper guide for human affairs, there is no sense that the natural world might be permanently disrupted by human activity. It is true that natural disasters, as well as apparent disruptions of nature, such as eclipses, were interpreted as due to human misconduct. However these were the means by which heaven conveyed its disapproval. Disturbances were not due to direct action by humans on the environment but to moral lapses, particularly too much sexual activity or other self-indulgence. It was never assumed that heaven's displeasure would be permanent; sooner or later, the natural order would be restored.

To the extent that many passages in the *Dao De Jing* imply a philosophy of quietism, simplicity, and noninterference, this work presents an ethic which is consistent with attitudes held by many in the modern ecological movement. We must add, however, that it does not advocate anything like environmental activism. Most important for our present inquiry, the notion of nature and Dao in the *Dao De Jing* does not include anything we could recognize as science. Though the Dao in Chinese thought has been likened to natural law, it is a sort of law accessible only for those with a special gift and which by its nature cannot be formulated in language. Central to all forms of Chinese philosophy is the notion of oneself, or the state as a whole, being in accord with the Dao or not in accord with it. Being out of accord with the Dao will adversely influence the well-being of both individual and state. When an evil emperor is on the throne, the Dao is absent from society, and the wise person withdraws, knowing the Dao, or natural order, will eventually restore itself. This is quite a different view than that of modern ecology, which fears permanent destruction of the earth's natural balance as a result of pollution, global warming, radiation accidents, or other harmful actions of humans. The most extreme fear is that the earth's ecosystem will be so damaged as to no longer support human life. The view of the *Dao De Jing*, while it warns of possible societal disaster, never doubts eventual restoration of peace.

## RETREAT TO THE COUNTRYSIDE

The *Zhuangzi* shares many ideas with the Laozi but has a slightly different sensibility. It repeatedly emphasizes the notion that only those who do not want to rule are fit to do so, because they are not seeking power. Many of its anecdotes concern a person refusing high office in favor of caring for himself. This reflects the very real hazards of government service in an era when an office holder who feels out of favor or was discredited by treachery became vulnerable to exile, mutilation or execution. Rural life far from the center of power was thus healthier, but the health hazards to be avoided were socially determined ones, not pollution or other urban blights.

The Daoism expounded by Laozi and Zhuangzi developed into a sort of eremeticism that involved removing oneself from the capital to retreat into the country where one idled away the days in drinking while enjoying the four pursuits of calligraphy painting, chess, and qin playing. Paintings often show a small group of friends sitting in a natural setting enjoying these pursuits. These activities were much commemorated in poetry and painting. For most, this sort of life was probably out of reach except by daydream, yet it was a vision that held the imagination of educated men throughout Chinese history. While avoiding politics was, as already pointed out, a motive for withdrawing to the country, there was also the belief, common in the modern West, that life away from cities was more wholesome. In this sense, it is akin to the modern romantic pastoral movement, but antedating it by millennia.

## ANIMAL WELFARE IN ASIAN RELIGION

Here is one of the most famous passages in Chinese philosophy, from Zhuangzi:

Chuang Tzu and Hui Tzu were strolling along the dam of the Hao River when Chuang Tzu said, "See how the minnows come out and dart around where they please? That's what fish really enjoy!"

Hui Tzu said, "You're not a fish – how do you know what fish enjoy?"

Chuang Tzu said, "You're not I, so how do you know I don't know what fish enjoy?" (Watson 1968:188f)

This entertaining anecdote that I have truncated somewhat, has evoked commentary far out of proportion to its modest length. Philosophers have tended to approach it as posing an epistemological question: Can we know what others are thinking and feeling? A more straightforward implication is that fish are like us in that they are capable feeling enjoyment. Chuang Tzu simply sets aside his opponent's arguments because he is confident in his own perception of fish emotion. This passage is not, of course, evidence as to whether fish feel emotion. What is relevant for our present inquiry is Zhuangzi's intuition that other creatures are like us in their ability to feel. This way of thought is at the heart of the modern animal rights movement. Sadly, there is little reason to think that this passage, known to nearly all Chinese literati, made any difference in how animals were treated in China. As in other agricultural societies animals were employed for labor and food without any apparent sentimentality; indeed, human survival depended on their exploitation. Mistreatment of animals in the developed world continues but is kept out of sight.

Exploitation of animals can take odd forms. In China the shamanistic practice of assigning human traits to animals and then ingesting them to enhance such traits has only increased with recent prosperity. Wild animals are caught and sold in markets at high prices to gratify the superstitious belief that eating their flesh will somehow strengthen health. The SARS epidemic that killed many thousands of humans was spread from civet cats caught in the wild and sold in Southern Chinese markets. It is easy to single out Chinese for this sort of behavior simply because it still exists. We should not imagine that other traditional cultures were necessarily more concerned with animal welfare. Vegetarianism was required in India for the Brahman caste and cows were sacred, indicating a greater degree of concern for animal welfare, though this was incomplete.

It was Buddhism, not Daoism or Confucianism, that brought vegetarianism to China. Shakyamuni, the historical Buddha, forbade monks consumption of meat when the animal had been killed specifically for them; complete avoidance of animal foods was a later refinement of doctrine. It is most prominent in East Asian Mahayana, being obligatory for monastics but not usually practiced by the laity. The reasons for the proscription of meat in the Mahayana are several. First, a meat-free diet was supposed to help subdue the passions. This is the other side of the belief that meat enhances virility. For monks and nuns, not only meat is banned but also onions and garlic; that these vegetables inflame the passions is an ancient Indian belief. The other important rationale for veganism was the doctrine of karma and rebirth. It was commonly stated that since sentient beings have as many rebirths as there are grains of sand on the banks of the Ganges, each living creature has been one's mother, father, brother, sister, husband, wife, etc., in a previous life. Thus to eat an animal is to risk eating those who were most dear to you in a previous life.

Viewed metaphorically, the notion that animals were human in earlier rebirths and that we might someday be reborn as animals indicates a sense of commonality with all of life. In this it is like Zhuangzi's fish story. The doctrine of rebirth had only a limited positive effect on treatment of animals. As so often with religious teachings, the Buddhist admonition against cruelty to animals became ritualized. At certain seasons, captured animals were purchased to be released, an act thought to earn merit for those who released them. Accumulation of merit was subtracted from one's burden of unfavorable karma. While in Cambodia, my wife and I participated in such a ritual. When in the vicinity of a temple, we were approached by a woman carrying a wooden cage of small birds. One pays so much per bird, takes them in hand and releases them; soon they fly back to their owner for a secure meal and to await the next merit seeker. This is a charming custom and probably a good deal for the birds, but this custom is far from the original intent of freeing animals destined for the dinner table.

## THE ENVIRONMENT IN CHINESE ART

One of the best ways to gain a sense of what nature meant to educated Chinese is to look at Chinese landscape painting (Yang, et al. 1997). This genre of painting began in the Tang and lasted through the Qing, with considerable stylistic evolution in between. To the inexperienced eye, such painting may seem repetitive but to those who understand the ideas underlying it, the variations exercise a subtle fascination. Except in official portraits, humans are not depicted in isolation but in their environment, nearly always a natural one. It has often been pointed out that in contrast to European painting, that of China always shows humans as part of the larger world. The Chinese term translated as "landscape" is actually "shan shui" meaning "mountains and water." Virtually all such paintings have both elements in order to provide a balance of yin and yang. The real subject of such paintings is the harmony of nature. Humans and their dwellings are always quite small in proportion to the natural features and there is always a path that the viewer can imagine him or herself treading. This harmony is built up from brushstrokes that are always quite visible and bear some similarity to calligraphy. The effect is far from the realism of a photograph; rather the art is to make nature and the brushstrokes one and the same. Consistent with the Buddhist doctrine of rebirth, animals are depicted as sentient beings, not objects. Some animals, especially fish and tigers are commonly shown with an expression suggesting the wisdom of enlightenment. In contrast to Western illustration, such as that of James Audubon, the animals do not look like objects but convey a sense of being conscious.

## VIRTUOUS DEFORESTATION

Through its art and Daoist philosophy, China expressed its vision of humanity as part of nature, not its master. Nonetheless there was fear of uncontrolled nature in China as in all preindustrial societies. Deforestation, for example, was seen as beneficial for the human environment. To understand this we must remember how different life in early China was from our own time. First, most of the country was forested and thus unavailable for cultivation. Clearing land increased what was available for growing food in an era when starvation was always looming. It also diminished the threat of attacks by wild animals such as tigers. With early technology that provided only plows and crude axes, there was no worry about excessive clearing. Eventually as deforestation continued, it caused harm to the water supply. With fewer trees, water flowed down more rapidly as snow melted in the spring. As a result there was flooding early and drought later.

The leading scholar of Chinese ecology, Mark Elvin, summarized the ambivalence toward uncontrolled nature as follows:

. . . classical Chinese culture was as hostile to forests as it was fond of individual trees. (Elvin 2004:xvii)

We find this attitude in two well-known passages from Mencius. The first talks of trees but, as so common in Chinese literature, as a metaphor for something else:

Once the trees on Ox Mountain were quite beautiful, but because it is on the border of a large state, they were continuously chopped down with axes. (Modified from Lau 2003:251)

This passage has appealed to many because Mencius seems to be expressing concern regarding destruction of nature. Reading farther, however, we discover that Mencius is simply using Ox Mountain as an analogy for a person's failure to cultivate himself:

The way in which a man loses his proper goodness of mind is like the way in which trees are denuded by axes and bills (Legge 1892:408).

Our expectation that this passage will conclude with an admonition against despoiling nature is disappointed. Instead Mencius turns the loss of beauty into an analogy for neglect of ethical self-cultivation. As we have seen in so many other instances, Chinese observation of nature is turned into an analogy for an ethical lesson. Here, as is usually the case, the implications of the observation were not developed beyond the analogy.

Here is another passage from Mencius, this one expressing the widespread Chinese fear of untamed nature:

> In the time of Yao [a mythical ruler], when the world had not yet been perfectly reduced to order...Vegetation was luxuriant, and birds and beasts swarmed. The various kinds of grain could not be grown. The birds and beasts pressed upon man...measures to regulate the disorder were set forth.... Yi set fire to, and consumed, *the forests and vegetation on* the mountains...so that the birds and beasts fled away.... When this was done, it became possible for the people of the Middle Kingdom to...get food for themselves. (Legge 1895:250f)

Although the specific events are mythical, the passage unambiguously praises destruction of the natural environment as a way to get rid of the animals that live there. Elvin notes that habitation on cleared land, rather than living in the forest, was considered an essential trait of being Chinese, that is of being civilized (Elvin 2004:42–44). An attitude farther from that of contemporary ecology can hardly be imagined. Yet we must bear in mind that at the dawn of human civilization most of the earth was covered by dense forest that could not be cultivated. Wild animals, rather than being appreciated for their aesthetic value threatened human life by destroying crops or by predation, either of domestic animals needed for food and labor or by literally eating people. The mixed effects of forest on human life are well described by Elvin who observes that, "...we are forced to confront the ambiguity of forests" (Ibid. 35). Many passages in Chinese literature express the fear travelers felt upon entering forests, which harbored not only bandits but also hungry tigers.

In a sense, the use of natural phenomena as metaphor gave these phenomena spiritual significance, as in the Mencius passage quoted above in which failure to protect a spot of natural beauty is like a person's failure to cultivate his or her virtue. (The English term "cultivate" is itself an agricultural metaphor.) Nature thus had several aspects, to some degree contradictory. On the one hand, it was the earth of the heaven-earth-man triad and it was where one retreated, if only in imagination, to escape politics as well as a source of analogies for philosophical discourse. On the other hand, nature had to be controlled for civilization to survive.

Lack of concern for forests as resources did not end with the fall of the imperial system in the early twentieth century. A major contributor to deforestation was a bizarre social experiment during Mao Zedong's Great Leap Forward. To show the power of the masses, steel was supposed to be produced by rural farmers in backyard blast furnaces. This produced almost no usable steel but consumed much wood, diverted labor that was needed for agriculture.

## CHINESE ECONOMICS: FEAR OF SCARCITY

At times we do sense something akin to modern ecological awareness but it typically turns away from real issues to mythical ones. Here is a passage from the *Huainanzi:*

When the age of decadence arrived, people cut rocks from mountains . . . extracted pearls from oysters, and smelted copper and iron ores. After this the multitude of living things multiplied no longer.

They slit open the wombs of wild animals, and they butchered their young, after which the unicorn roamed around no more. They tipped over the nests of birds and smashed the eggs, after which the phoenix ceased to soar aloft.

. . .

Even with less than the full number of people and full amount of equipment, their storehouses still had a surplus. But the multitude of living things did not proliferate. The greater of them perished as sprouts, or in the egg, or womb.

. . .

The heavens dried up and the earth split in fissures. The phoenix descended no more, and the wild creatures with talons, fangs, and horns who sought to escape were thus impelled to confused violence. The common people . . . froze and starved, the dead lying on each other. (Elvin 2004:109–110)

Elvin summarizes this fascinating passage as, "an evocation of the most critical shock in human history: the onset of progress." It resembles modern ecological concerns to a striking degree, particularly in its fear of technology and in the paradoxical statement that despite a surplus of food, living things were becoming extinct. It is the ancient Chinese version of silent spring. But there is a big difference; in the Chinese version of ecological desolation, the most prominent loss is mythical; the vanishing species are the unicorn and phoenix. Once again it is not so much the direct actions of humans—those cited such as slitting open wombs seem fanciful—but the disturbance of the heaven-decreed pattern of human life that results in disaster. This myth that lack of virtue will result in the countryside's being unable to sustain life is found in many cultures and in modern environmentalism as well. That environmentalism has parallels in ancient myth does not mean that present day threats are unreal, simply that fears of this sort are deeply ingrained in the human mind.

The elite also had practical concerns about the adequacy of the food supply. Worry about agricultural productivity did not necessarily imply solicitude for the welfare of farmers. Instead there was apprehension that the farms would be deserted for easier forms of work. Ideally, according to the Warring States text *Guanzi Jiping*, farmers would:

Exhaust the energy of their four limbs to pursue unremittingly their work in the fields...

...

They will have been accustomed to this since they were young, and their hearts will be content with it. They will not be enticed into going off elsewhere by the sight of unfamiliar things. (Elvin 2004:105)

This seems a decidedly unspiritual attitude toward the miseries of agricultural labor but something like it is to be found in the rather notorious Chapter Three of the *Dao De Jing*, as indicated in the following excerpt:

Not seeing desirable things prevents confusion of the heart. The wise therefore rule by emptying hearts and stuffing bellies, by weakening ambitions and strengthening bones. (Feng and English 1972)

This shows the shadow side to extolling the simple life: It may be a pretext to keep those with nearly nothing from aspiring to something better. (Many who admire the philosophy of Laozi, including the present author, find this passage inconsistent with the overall tone of the work. Clearly the *Dao De Jing* is a composite and so there is not complete consistency within the work. This particular section has often been attributed to the influence of the legalist school, which advocated strict laws and punishments as the best means to maintain social order. While this is plausible, we must guard against altering ancient texts to better fit our own ideas. We must attempt to read them for what they do say not for what we want them to say.)

Fear of scarcity was perennial in China. The economic assumption was what is now referred to as Malthusian: There was only so much of everything, particularly food; there would be starvation if the supply were not carefully managed. This assumption was true for early civilization but long-term food adequacy can never be assured by rationing, increased productivity is necessary. There were periods of starvation in China, just as there now are in Africa. In most cases famine is caused by poor management, for example, lack of flood control, not by inadequate resources. Political ideology was a factor at times, for example in the notorious "Great Leap Forward" of Mao Zedong. In the modern west, there is fear of scarcity too, but of energy rather than food.

China had more than enough land, though much was not arable; the limiting factor, at least in the government's view, was labor. In early civilizations, with life expectancy being considerably less than thirty, natural resources were seemingly unlimited, but there were not enough people to cultivate them, even in China. This was a salient difference from the modern world in which the fear is that population may grow in excess of resources. Eventually, as the Chinese population grew it was land rather than labor that was seen as limiting (Elvin 2004:181–186).

Here are passages from a Warring States text, the *Guanzi Jiping* which discuss agricultural and resource issues from the perspective of practical government policy rather than philosophy:

Even if the pools and wetlands spread far and wide, and the fish and the turtles are plentiful, there must be regulations on the use of nets. . . . How can the people be allowed to stop producing grain? This is why it is said that the kings of former times put prohibitions on activities in the hills and wetlands *in order to increase the involvement of the people in the production of grain.* (Emphasis in original translation.) (Elvin 2004:103–110)

As this passage begins, it seems to express a concern about depletion of fish by overfishing. In fact, however, the concern is that if farmers can more easily obtain food with a net, they will be able to avoid farming.

The ruling class was well aware that agricultural labor was grueling:

Energy is produced by exhausting our bodies. The reason that the ruler has limitless resources is because the common people exert themselves without respite. (Ibid. 103)

This view is an inversion of contemporary values. That wealth is created by the exploitation of the workers is admitted quite openly, but without any apparent moral qualms. Rather, overworking laborers is necessary and appropriate to support the lavish lifestyle of the imperial court. There is no conception of social or economic progress, of better lives made possible by improved production methods. This is not to say that no officials cared about the welfare of the people, many did. However such concerns were not broadly recognized. Confucianism was a virtue ethic, speaking of benevolence, reverence, and filial piety but offering few specifics. The Daoism of Laozi simply recommended a way of life modeled on that of the ancient past. While there were relief efforts, coherent long-term social policies for relieving the misery of the ordinary people were underdeveloped. To a great degree, their welfare depended on the character of individual officials who were rotated frequently to limit their involvement with their subjects.

As Elvin points out, the military was particularly concerned that agricultural production be maintained, presumably so that soldiers could be adequately fed. Another significant point here in relation to technology is that, at least in this early text, the only way to assure adequate food production was keeping enough farmers alive while working them without respite. There is no apparent awareness that systematic study of agricultural methods might result in improved productivity of both land and labor. While later texts do urge improved agricultural management, especially with regard to water for irrigation, this was more a matter of applying existing methods than invention of new ones. While the modern concept of progress—that innovation can improve quality of life—is easy to criticize, without progress, there is little chance for human life to improve outside of the privileged classes, whose wealth is the result of the labor of others.

If the government did not consistently try to stimulate innovation, it did recognize the need to promote efficient agriculture:

It is the business of the forest wardens to se that the rules about fires are enforced, so respect is shown to the hills, wetlands, forests and thick vegetation. This is because these are the sources of materials. Prohibitions should be imposed and lifted in the appropriate seasons so that the people have materials for their houses and stockpiles of firewood . . .

It is the business of the Director of Farming to appraise the lie of the land, to judge which soils are more fertile and . . . to consider what is suited to the terrain . . .

It is clear from these passages that even in early times, Chinese governments were concerned about resources, though it is not clear how effective their management was or to what degree their ideas about farming were empirically based.

## FENG SHUI AND ECOLOGY

Feng shui, sometimes termed "geomancy" by Western authors but literally "wind and water" is yet another aspect of Asian culture that has become faddish in the West. As with other such adaptations, modern Western accounts, heavily influenced by New Age ideas, bear only a partial similarity to the actual phenomenon in its country of origin. In particular, feng shui has been modified to render it a form of environmentalism, with a heavy measure of aestheticism. The latter is indicated by the tendency of bookstores to place books on feng shui in the home decorating section, rather than with books on Eastern religion. As we shall see, traditional feng shui has little to do with contemporary concerns regarding the environment or home décor.

Selection of the proper site for the homes of the living and the tombs of the dead seems to have been of particular importance for the Chinese from at least the Shang dynasty, though the actual origins of feng shui remain uncertain. It is closely associated with so-called "ancestor worship," actually a complex of beliefs and customs regulating the relationship of the living to their deceased ancestors. It was always assumed that these must be propitiated to ensure their aid and, what received the main emphasis, in order to prevent them from harming the living. Fear of ghosts remains active among many Chinese. Only a few decades ago, a successful cinema in Hong Kong was forced to close because of rumors of ghosts in the washroom. For modern urban or suburban dwellers, ghosts are more for entertainment than fear but anyone who has been lost in a strange city or in a forest will recognize the irrational fear such experiences evoke. Confucius, who famously declared that he kept his distance from spirits, recast rites for the dead in terms of filial piety so that ancestor devotion could be seen as ethical rather than supernatural. In actual practice these explanations were complementary.

As already explained, feng shui had two functions: selection of sites for the living—homes and businesses—and grave siting. The latter has always been particularly important because of the generally unquestioned belief that unsuitable graves, for example, those whose line of sight was blocked by tall trees, would engender malice of the dead toward the living. Grave selection could lead to family disputes because a site that would favor one family member might be harmful for another. At times, the dead did not receive final burial for many years while they awaited resolution of disputes about the optimal location according to feng shui.

Some of the principles of feng shui are mentioned in Warring States and Han sources. The *Book of Rites* (*Liji*), probably compiled during the Latter Han (Lowe 1993:293–295) mentions that the heads of cadavers were placed toward the north, this being the direction associated with yin and, therefore, death. Separation of the dead was a matter of great concern to a culture that assumed the deceased could affect the fortunes of the living (Bruun 2003:273). Homes were to face south and etiquette always dictated that the person of higher rank faced south and the inferior one north.

Feng shui is divided into two forms, though in practice each used techniques associated with the other (Bruun: 279). The school of forms and outlines (*jing shi*) emphasized the shape of the terrain while the school of directions and positions (*jang wei*), also referred to as the compass school, gave a central role to the *luopan*, or geomantic compass. This latter school utilized many of the terms and symbols of orthodox cosmology such as the *bagua* (eight trigrams). No doubt, within each school there was considerable variation between the actual methods of individual practitioners. The luopan is constructed on cosmological principles with its base in the form of a square, the traditional symbolic shape of earth, and a rotating circular disk, symbolizing heaven. A compass is set in the center to permit precise orientation. On the disk are varying numbers of concentric circles marked with standard cosmological icons including the five phases,

*Yi Jing* hexagrams, constellations, the stems and branches used for dates, various heavenly mansions (sky segments), and fortunate and unfortunate locations (Feuchtwang 2002:297–299 and passim). The movements of the compass were easily attributed to qi since the force that produced them were invisible. The compass was clearly a Chinese invention and the fish-shape of the first one described outside of China in an Arabic text suggests Chinese origin (Temple 2002:149–157). Textual sources indicate that the compass was used for navigation but its most pervasive use was for feng shui. Luopan are readily available in bookshops and other emporia in China, Hong Kong, Taiwan, and Western Chinatowns, but are extremely expensive in relation to their materials and workmanship. This suggests that they have high value for those who use them as professional feng shui masters and, possibly, for collectors. The compasses mounted on them are of surprisingly poor quality and do not at all point in exactly the same direction, suggesting that their users are not meticulous about directional accuracy.

The luopan possesses a definite mystique and must have greatly impressed the clients of feng shui practitioners. This quasi-magical allure of scientific instruments persists to this day in such devices as the stethoscope. The principles of use of the luopan seem quite difficult; many of the characters inscribed on the rings denote phrases such as "white dew," "valley rain," and "little fullness" that cannot be understood by most Chinese. However the luopan does have standard methods of use, which can be acquired through study (Walters 1991:165–182).

The other school of feng shui bases its judgments on the appearance of terrain and in this respect might seem closer to modern Western versions that graft aesthetic concerns onto feng shui. However, the function of feng shui of both varieties to the Chinese is simply to assure good fortune. An example is the standard recommendation that living things be placed in a home or business. Houseplants, because they are alive, are beneficial. The fish tanks, nearly always with seven fishes, usually carp, that are found in many Chinese homes and restaurants are there because they are considered auspicious, not because of any particular love of animals. The example of fish illustrates another aspect of Chinese symbolism: the use of homonyms, basically puns. The word for fish, *yu*, while written with a different character, has the same sound as "abundance" or "more than enough." Hence fish are assumed to facilitate accumulation of wealth. Some other homonymic associations are bats, *fu*, which is the same sound as good luck. Hence pictures of bats, often combined with other auspicious symbols, are often posted in homes and businesses. Lotus seeds, *lian zi*, rhymes with *lian sheng gui zi*, "consecutive births of noble sons," so lotus seeds are eaten at New Year in order to facilitate birth of sons. Several of these examples clearly have an aesthetic dimension and so they tend to be emphasized in presentations for Westerners. They make up a small part of traditional feng shui, however. Stressed-out Americans particularly want a home environment that is harmonious and soothing and so feng shui is presented as embodying these values. This is not to say that Chinese do not also value pleasant surroundings. That their luxuriant gardens and elegant ceramics are sufficient proof of this. Yet the function of feng shui was not aesthetic but practical, to facilitate the usual wishes of Chinese: wealth, health, long life, and male offspring. Chinese have no monopoly on these desires but are perhaps more willing to openly admit to them.

Ole Bruun (2003:269–270) emphasizes that in China, feng shui does not embody any sort of environmental ethic. He does acknowledge that in the past, the home compounds of the gentry had extensive foliage and that landscape painting also emphasizes this. But in contemporary China, on the basis of his extensive fieldwork, he notes that Chinese villagers will use anything natural to their advantage. Trees, when there are any, will be quickly chopped down for sale or for firewood. Birds are shot for food or fun and almost any animal provides a welcome addition to the dinner table. Such are simply facts of life for those who live in a subsistence economy. Here is how he summarizes the situation:

... wildlife simply does not exist after an entire century of civil war, recurring periods of starvation following disastrous political campaigns and, on top of all this, inadequate interest in preserving whatever species that may be left. Thus, to make it absolutely clear, feng shui does not entail making the earth a suitable living place for other creatures than humans and their domesticated animals. (Ibid. 250)

Bruun trenchantly points out that a Chinese villager, if successful in exploiting the environment, would likely interpret this personal benefit as due to the favorable feng shui of his home. Far from encouraging conservation, feng shui can be used as an evasion of it. This ancient system is so magnetic for Chinese that it fills the mental space that would otherwise be open to true environmental concern. Asking how to make a home or business healthy for both humans and the environment is an important question but feng shui takes the place of a more scientific approach. In the case of a business, gathering data such as proximity of competition, neighbors' need for the products sold, and so on, would seem more likely to ensure success than site selection by luopan. It could be argued that feng shui might produce something like a placebo effect, making Chinese feel more secure in their homes and businesses. Certainly the converse is true—many Chinese feel less secure if they feel their location has bad feng shui. Feng shui is such an important part of Chinese culture that it is unlikely to be abandoned anytime soon.

## THE ENVIRONMENT IN CONTEMPORARY CHINA

No visitor to China can escape awareness of the pervasive pollution, typical of developing countries, the result of putting economic development before health. Thus whatever compatibility there may be between the Chinese regard for the natural harmony of the Dao and modern environmentalism, it has not had a noticeable practical impact. At this time, China is second only to the United States in emissions of greenhouse gases and is projected to attain first place within a decade or two. Since China is only at the beginning of its phase of accelerating economic development, consumption of energy and production of pollution will increase far beyond that of the present developed countries unless conservation is given more attention. Already there are beginnings of awareness that unabated pollution is harmful and also that it diminishes the country's stature in the developed world.

With respect to the environment both science and China's spiritual traditions have failed to stem the effects of carelessness and greed. Yet surely it is in protection of the earth that science and spirituality are both needed, the first for the necessary technical knowledge, the second for ethical principles. It seems certain that adaptation of modern conservation by the Chinese government will be inspired less by the philosophy of Laozi and Zhuangzi than by objective scientific evidence of the harm unbridled pollution will inflict upon China's people. The Danish anthropologist Ole Bruun, on the basis of his extensive fieldwork in China, paints a grim picture of the current condition of its natural environment. Uncontrolled use of insecticides has not only thinned the insect population dramatically but also that of the birds that fed upon them. Songbirds are rarely heard and villagers do not seem to miss this simple pleasure (Bruun 2003:250–251). Efforts to educate villagers in the care of the environment have little effect. Those supposed to go to the countryside to train farmers in conservation find ways to avoid doing so. Edifying messages beamed from loudspeakers are, understandably, tuned out by the villagers.

Needham terms feng shui a "pseudoscience" and it is difficult to dispute his characterization. Geomancy, like its heavenly cousin astrology, borrowed scientific devices such as the compass and makes use of extremely elaborate calculations (Koh 2003). This hocus-pocus gives it the trappings of science but not the substance. Yet feng shui contains a self-evident truth, that location of buildings can profoundly affect human well-being. Concern for grave location, while primarily driven by fear of ghosts, presumably included the human desire to treat the deceased with respect. Such concerns are not lacking in the modern West. At the time of this writing, rebuilding the New York World Trade Center continues to be held up, nearly five years after its tragic destruction, by disputes as to the proper memorial to its dead whose body parts remain in the crater.

Though the principles of traditional feng shui are quite unlike those of the modern ecological movement, the tradition can, it is hoped, incorporate new ways of thought. As shown in its Western adaptations, concern with a harmonious home environment is easily incorporated into feng shui. Thus houseplants and fish can continue to be recommended but for calm rather than for good luck. If feng shui can set aside its tree-cutting doctrines, one can hope it can turn into a force for conservation. This would not be the traditional feng shui but one more suited for modern life. There is ground for hope. Taiwan, an island with many beautiful natural features, as well as ugly factories, has a strong environmental movement and one hopes this will spread to China itself.

# Chapter 8

## Medicine in Traditional China and India

Interest in Chinese traditional medicine has blossomed in the West since the normalization of relations by President Richard Nixon and Chairman Mao Zedong in 1972. *New York Times* columnist James Reston stimulated popular Western fascination with Chinese medicine when he reported on his appendectomy in Beijing under acupuncture anesthesia. Once only practiced surreptitiously in Chinatowns, acupuncture has gone mainstream in the West with accredited degree programs and licensure available in many states. Although Chinese herbal medicines are used by Westerners, their use is casual, not based on the elaborate cosmologically based diagnostic and pharmacological system of Chinese tradition. Then too, herbs lack the exotic appeal of acupuncture, as much a symbol of China as chopsticks and pandas. Since the West has its own herbal medicine industry, those from China have significant competition. Also limiting acceptance is concern about purity; many are adulterated with heavy metals such as lead, arsenic, and mercury. A Chinese herbal treatment for prostate cancer was banned by the U.S. Food and Drug Administration because it contained the rodenticide warfarin.

Ironically, the trend in the People's Republic of China has been away from acupuncture toward herbal medicine. During the Mao era acupuncture was officially promoted, most likely because of its organizational simplicity and low cost. Now herbal medicine is being encouraged because it is more lucrative for hospitals. In both cases, political and economic motives, not safety or efficacy, determined policy. Traditional Indian medicine, *Ayurveda*, has also established itself in the West, not because of any dramatic event comparable to James Reston's appendectomy, but stimulated by the attraction many feel for Indian spiritual practices such as vegetarianism, Yoga, and mantra recitation.

Interest in Chinese and Indian traditional medicine today takes two forms. Some seek therapeutic benefit they have not obtained for diseases such as arthritis and cancer. Even more are apprehensive about side effects from mainstream medicine, which is widely feared by the public, despite its spectacular success. Some are drawn to it because of devotion to an Asian spiritual practice.

A different reason for interest in traditional medicine is the desire to understand the very different ways other cultures have conceived disease and healing. Here the systems of China and India are of particular interest because their practice today is in continuity with their origins at the dawn of human civilization. Medicine is the only traditional "science" still in widespread use. Many other traditional practices, especially divination, persist but they are not in any way related to science. For all but a minority of Westerners, divination is dubious while healing is something real.

## MEDICINE AND HEALING IN CHINA

Defining just what is to be classified as medicine in traditional cultures is not self-evident because medicine as a rational discipline was never entirely separated from other practices, such as exorcism, ritual, and use of amulets or talismans. Not everything done on behalf of a sick person is medicine. For the sake of clarity, in what follows, I generally use the term "healing" to refer to any means of relieving disease, however defined, and "medicine" to refer to use of drugs, herbs, surgery, dietary change, or other means intended to correct bodily dysfunction by nonsupernatural means. Thus religious practices such as offerings to atone for past life misdeeds are part of healing but not medicine. In classifying a modality as medical, I am not implying anything about efficacy. The question of the effectiveness of traditional medicine is a complex one about which opinions abound but data is lacking.

Notions of disease in the past were quite different from those of modern ones. Putative causes of disease included spirit or demon possession; curses including placement of malevolent charms in a victim's home; moral impropriety in this or previous lives; failure to carry out rituals for the dead; exposure to cold, heat, and wind; improper diet; sexual excess; and alterations of internal energies, usually described in terms of yin-yang or the three *doshas*. From a modern view, diet and exposure seem plausible in principle as causes of disease, but such nonverifiable events such as past life misdeeds are not. However for the most part, these kinds of causes were not separated in China or India. Some practitioners were more skeptical and rational but there was no absolute distinction such as we would make between antibiotic therapy for pneumonia and expiatory rituals for the same condition.

Within the extremely complex formal system of Chinese etiology and diagnosis, there was also a simplified version, often the basis of everyday practice, that attributed disease to heat or cold. This survives even today. It makes use of herbs that are classified into categories such as: those that clear heat, those that warm the interior, those that drain downwards, those that expel wind-dampness, those that regulate the qi, and so on (Yang 2002). In hot–cold theory, the cause of disease is assumed to be harmful actions of environmental factors on bodily functioning. The emphasis on heat and dampness is inevitably evident when one bears in mind that before central heating and air conditioning the discomforts of winter and summer were inescapable. Hot–cold theories of disease exist in many cultures; often modern pharmaceuticals are fitted into these schemes, so that drugs are considered either warming or cooling. A vestige of these theories in the West is the word "cold" for an upper respiratory infection even though the cause is not exposure to cold temperature but infection by a virus.

Though much in scientific medicine is established beyond doubt, there are always issues that are hotly contested. Hormone therapy for menopause and proper use of stents for coronary heart disease are current examples. For those of us caught up in these clashes of opinions, it is amusing to note that in traditional China there were equally intense controversies. An example of a contested area was the relative importance of hot and cold in disease causation. Some physicians held that all ailments were due to cold, while others implicated heat. That no factual information existed to support either position did little to dampen the debate.

Traditional medicine was not monolithic. There were competing theories and dissatisfactions just as there are within medicine today. Paul Unschuld points out:

The history . . . is not characterized by simple linear succession, in which practitioners exchanged each old system for a new one. Instead, the evidence reveals a diversity of concepts extending for more than two thousand years. New ideas were developed, or introduced from outside and adapted by authors of medical texts, while at the same time, older views continued to have their practitioners and clients. (Unschuld 1985:5)

Nor was there universal satisfaction with the existing system:

. . . those thinkers seeking solutions to medical questions wandered aimlessly in all directions, lacking any orientation, and unable to find a feasible way out. (Unschuld 1985:197)

Contrary to the currently fashionable nostalgia for prescientific medicine, earlier systems seemed to have no more fully satisfied practitioners or patients than does the present one. Unschuld notes that one's response to the progressively increasing dissatisfaction was to go back to pre-Song sources. This was the standard reaction to any sort of social difficulty—to try to return to a mythical past when everything did not work properly. This same atavistic attitude is present today in the desire to return to traditional healing.

## WAS TRADITIONAL MEDICINE "HOLISTIC?"

There are two widespread misunderstandings about traditional medicine that need to be corrected. First, it is common for modern interpreters of traditional medicine to laud it for avoiding mind-body dualism. While it is true that mind and body were not distinguished in Chinese philosophy as they are in Western science and philosophy, this was because the issue was simply not recognized. The relation of mind and body continues to be an intractable problem in both philosophy and neuroscience. Ignoring the difference is not a solution; what is to be hoped for is understanding of the mechanism by which mind and body interact. The philosophical notion that dualism is somehow an inferior way of thought to monism has no relevance to medicine. Nor does it have to science generally, which has many principles, not one. (The as yet undiscovered "theory of everything" in physics would not explain everything, but only a few conundrums in theoretical physics.) The notion that holism is desirable in medicine is a fantasy. In medicine precise location of the origin of symptoms can be a matter of life and death. For example, if you are having palpitations—the sense of one's heart beating too hard—it will make a great difference whether the cause is anxiety or a potentially fatal electrical dysfunction in the heart.

A related claim is that scientific medicine fails to treat the "whole body," while Chinese and Indian medicine somehow do this. The notion of "holistic" medicine, though it has entered popular culture, resists clear definition and so is little more than a slogan. If, for example, you have abdominal pain, the proper treatment depends very much upon localized diagnosis. It matters greatly if the problem is appendicitis, a kidney stone or an inflamed gall bladder, for example. As to treating the whole body, both Chinese and scientific medicine will identify a specific organ as the site of disease, although it will not be the same organ because the symptoms associated with specific organs are entirely different in the two systems Similarly, Ayurveda assigns disease etiology to changes in organs; the Indian conceptions of the organs differ from those of both Chinese and Western medicine. In Ayurveda, a usual way of deciding which organ is involved is to inspect the tongue, onto which the internal anatomy of chest and abdomen are mapped. The heart and digestive organs are on the midline, with heart near the tip and colon at the base. The lungs and kidney are on either side. A change in appearance of the tongue in a particular spot indicates that the organ correlated with that location is the site of the illness (Lad 1984:60–63). This is a form of correlative metaphysics in which the tongue is a microcosm of the entire body. The idea that abnormalities elsewhere in the body express themselves on the tongue can be called holistic but it is only imagery, no empirical evidence supports these associations. Arguably, procedures such as tongue diagnosis in which only part of the body is used for diagnosis is *less* holistic than Western medicine, which looks at all areas in the search for disease etiology. Diagnosis by tongue, or face, or pulse (as in the Chinese system) is really more akin to divination, such as inspecting patterns in tortoise shells or bird flights, than to scientific diagnosis.

These points are made not to disparage traditional diagnosis but to clear the way to studying it dispassionately, without projecting onto it the mystique of the exotic. There is no emergent property that makes a particular system of medicine holistic. A related criticism of modern medicine, that it focuses on the physical disease and does not address the other needs of the patient, is often true and deplorable, but is a practical issue, not a metaphysical one. It is by no means clear that medicine in traditional Asia did a better job of addressing patients' psychosocial needs. Claims to the contrary usually have underlying agendas, most often they are self-promotion by alternative practitioners. While the current Western system is in great need of constructive criticism, this is more likely to be effective if done directly, without comparison to prescientific systems.

## THE BEGINNINGS OF MEDICINE

Possession and spite expressed through curses seems to have been the earliest notions regarding causation of disease, at least in the oldest cultures for which we have records. These embody a natural human fear of supernatural agency and of being envied or hated. Ethical explanations in which the patient's behavior caused the disease were the next stage of development, emerging with the growth of moral sensibility. This form of explanation was added to the earlier notions rather than replacing them. Usually the diagnosis was made by a shaman or other specially gifted healer. Later as philosophical theories began to appear, rational theories developed based on correlative metaphysics. Evidence-based diagnosis and treatment developed sporadically and did not supplant correlative ones until the arrival of Western medicine.

Criteria for deciding which of these various etiological systems to apply were never clearly defined. The decision as to what sort of healing specialist to employ was fundamental and no doubt choice was affected by such factors as where the family lived, their economic status, their educational level, and their personal belief system. How patients decided between purely religious modes of treatment using exorcism or expiation versus rational ones employing herbs, acupuncture or lifestyle change cannot now be reconstructed.

Ethical theories of disease causation, while hardly scientific, did shift focus from external supernatural agents to the patient himself or herself. Ethical etiology places the blame on the victim and provides the opportunity for the healer to manipulate the patient by evoking fears of supernatural punishment. Within China, the ethical lapses thought to cause disease were usually misdeeds toward another or failure to follow rules of conduct. Sexual excess was commonly blamed; the treatment was a period of abstinence. As to how often such advice was followed by affluent Chinese men with a wife and several concubines, one can only speculate. In India too, depletion of semen was considered dangerously debilitating. Because women could produce a seemingly unlimited quantity of secretions during intercourse, it was assumed that frequent sexual activity was less harmful for them. This is sexology defined by men and should be interpreted accordingly. It is related to the common mythical notion of female insatiability and indicates that concern about sexuality and health was mainly, though not entirely, for the benefit of men.

This explanation of disease on the basis of sexuality continues to thrive in the modern world, notably in Freudian psychoanalysis that simply reverses the causality. Sex is still a cause of illness, both mental and psychosomatic, but now what is dangerous is not excess but frustration. These repression theories are no longer taken seriously in the mainstream of psychotherapy but they persist in pop psychology. Given the emotional and physical intensity of sexuality and its inherent risks, from hurt feelings to jealousy, to family disruption, to physical abuse and the risk of deadly infectious disease, apprehension about sexuality and health is inevitable.

In India, and later in China under the influence of Buddhism, karma from past lives was added to ethical causes of disease. Thus someone living a blameless life in this incarnation might have committed evil deeds in a past one. Within New Age ideology, the idea that people cause their own disease by improper attitudes has reemerged. Thus cancer has been held to be due to a "cancer personality" in which people never learned to talk about their problems, heart attacks are due to a type A personality, and so on. Research has refuted these sorts of causal theories, but many continue to subscribe to them. That character traits and behavior can cause ill health seems to be an archetype built into the human mind. Obviously there is truth in this belief. Morality and health do overlap in some conditions such as alcoholism and drug abuse, which tend to be considered both as moral defects and as unhealthy behavior. Those who do not exercise, who drink heavily, who smoke, or do not follow the latest notion of a healthy diet are often judged as lacking character; the tendency to associate morality with health remains strong, though the behaviors condemned have changed.

While theories of demonic or ethical causes of disease never died out completely, rational explanations gradually took the fore. These theories were posited as general principles from which diagnosis and treatment were deduced. They were not, for the most part, empirically based and there was no systematic observation of safety or efficacy. Etiology was explained in terms of hot–cold, qi, yin-yang, and the five phases in China and the three gunas and doshas in India. Treatment might involve dietary and lifestyle change, though in quite different ways than are in vogue today. Increasingly, treatment moved toward pharmacotherapy, that is, administration of substances thought to be active, including not only herbs but also such substances as fermented cow's urine and dried insects. Procedural treatments included acupuncture, moxibustion, and cupping. The unpleasantness of most treatments is striking and perhaps not entirely accidental. The association between something being good for you and distasteful or painful may also be built into the human mind, though why this should be so is puzzling. Fortunately, not all treatments were unpleasant; massage was popular, particularly in India.

The use of drugs and physical manipulations strikingly mimics modern medicine. The process by which treatments were adopted and maintained in medical practice remains mysterious; it was not based on actual evidence. There is evidence of efficacy for only a very few traditional treatments and evidence of harm for many more. Contemporary proponents of traditional medicine claim that long use verifies efficacy and safety, but this argument is fallacious. In the West, harmful treatments such as purgation and phlebotomy endured for millennia. Furthermore, the opportunity of learning from experience was much less because there were no journals or other methods of sharing results. Thus at least one Chinese physician recognized that if water is boiled before being consumed by a child it will not cause diarrhea. This discovery could have saved millions of children's lives but the poor organization of medicine and lack of an established public health system meant such discoveries were not universally disseminated. Lack of institutional structure for conserving and propagating scientific information as a major factor inhibiting scientific progress is discussed in detail by Huff (1993).

## TCM PAST AND PRESENT

What is now referred to as "traditional Chinese medicine" or TCM in the People's Republic of China is actually a modernized, somewhat simplified version of a three-thousand-year-old tradition (Eckman 1996:xii). An authoritative translation of an officially sanctioned TCM text is available (Sivin 1987). Like simplified Chinese characters, TCM was part of the program of the Chinese government under Mao Zedong that attempted to standardize Chinese culture in a form accessible to "the people." TCM is, therefore, despite its appellation, not entirely traditional. Indeed, the compression of the rich variety of Chinese medicine into a single official version was the product of the same denigration of the Confucian literati tradition as was simplification of the written language (Ji 2004). Political ideology was a primary motivation in the standardization of TCM; it does not appear that therapeutic efficacy was considered in developing the standardized form. Indeed, political ideologues would not have been capable of such an evaluation. The much-publicized "barefoot doctors," supposedly the answer to China's shortage of practitioners, were actually given only a few weeks' training and had no real diagnostic or therapeutic skill. This political control over medicine by the ruling Communist Party was nothing new but very much in the tradition of the imperial government's policy of controlling all kinds of knowledge. Since Western medications were, and still are, unaffordable by the overwhelming majority of people in developing countries, including China and India, an underlying government motivation for promoting traditional medicine has been to avoid unrest by giving the people the illusion that effective health care was available to them. The Communist government certainly did not want to expend the then very limited foreign exchange reserves on medication. From a government's perspective what matters is that the people are relatively satisfied with their health care, not that it is effective. Given the immense faith Chinese have in their traditional system, this approach was politically adroit, whatever its actual effect on the health of China.

## THE NATURE OF CHINESE MEDICINE

Paul Unschuld gives a useful sevenfold categorization of Chinese therapeutic systems: "oracular therapy, demonic medicine, religious healing, pragmatic drug therapy, Buddhist medicine, the medicine of systematic correspondence and, ultimately, modern Western medicine" (Unschuld 1985:4). It is what Unschuld terms "the medicine of systematic correspondences" that represented the mainstream in the elite Chinese medicine that gradually evolved into TCM.

The metaphysics of yin-yang and wu xing were the supposed theoretical basis of both acupuncture and botanical therapeutics. However, many of the specific therapies must have had other origins, particularly unwritten folk traditions, as well as actual observations by alert physicians. Any remedy could be given a place within the correlative system; once fitted in, its origin became unimportant and was forgotten. Conceivably, acupuncture may have begun with a simple observation that insertion of stone needles into a certain spot produced physiological changes such as alleviation of pain. Subsequently, a limited number of early points based on experience would be elaborated into a system of hundreds of point spaced along virtual meridians, most of which were undoubtedly invented to fill out the system without any empirical basis. Actual acupuncturists use only a small subset of the hundreds of points seen on the acupuncture dolls ubiquitous in Chinese shops. In practice, selection of points is probably partly empirically based, though the dogmas of revered teachers must also be a determining factor. The theory is not overtly altered to account for the implicit recognition that most of the points are useless because this would throw into question the entire metaphysical system and the authority of tradition.

Acupuncture thus may have started as a collection of practical methods, though there was probably religious motivation as well. As we have seen so often in Chinese science, rather than following up with systematic experimentation, further development was based on the superimposed correlative system, thus replacing empirical observation with scholastic reasoning. Texts established as authoritative, such as the *Lingshu* (Spiritual Pivot) (Wu 1993) also contributed to ossification of the system. Medical examinations were based on memorization of such texts and their attribution to ancient sages such as the Yellow Emperor placed them beyond criticism. That parts of these texts were not repudiated however does not mean they were always slavishly followed. No doubt physicians were selective; using some of the remedies and ignoring others, as is the case today. However the scientific process of a finding being announced, tested by others, and eventually accepted or refuted did not occur.

What we know of medicine of traditional China is unavoidably text based. Yet it was learned from teachers who must have had secret remedies that they held closely and only passed on to particularly favored students. We know something of the actual practice of medicine only anecdotally from literary sources (Unschuld 1998). Just as reading a contemporary medical textbook gives little idea of what will happen at a visit to the doctor, so actual practice in China must have had elements not derived from the canonical texts.

## EXORCISTIC AND SHAMANIC HEALING IN CHINA

It is likely that the text-based medicine was practiced by and for the affluent; what survives in textual form is the medicine of the educated elite. The majority of the population must have relied on healers with limited literacy whose knowledge of the theoretical basis was probably slight. Plausibly the less educated were more likely to patronize shamans, who employed oracular, demonic, and religious techniques. Daoist adepts frequently competed with shamans in offering such sorts of healing. A common remedy was to write a *fu*, a talisman against whatever inimical being, such as a demon or deceased ancestor, which was determined to be the cause of the illness. This was then burned and ingested by the patient. Such methods are mainly of interest to anthropologists; to moderns they lack the appeal of the text-based herbal medicine and acupuncture. Oracular healing is attested on many of the Shang oracle bones, whose inscriptions contain a query to the high ancestor god, Di, as to whether illness will come or, if it is present, what deceased relative had placed the curse. Treatment was by means of incantation. Therapy by means of drugs is not mentioned in the oracle bone inscriptions (Unschuld 1985:21).

In the Zhou dynasty, the spirit realm becomes a more complicated place, inhabited by demons who, in contrast to the Shang, were not necessarily relatives of the patient. Clearly it was in the interests of exorcists to play up the hazards, social as well as personal, imposed by the demonic realm. It was also in the interests of rulers to present themselves to the populace as the ones most capable of ritual control of these baleful influences. Here is a description of a public exorcism:

Several times a year . . . hordes of exorcists would race shrieking through the city streets, enter the courtyards and homes, thrusting their spears into the air, in an attempt to expel the evil creatures. Prisoners were dismembered outside all gates to the city, to serve both as a deterrent to the demons and as an indication of their fate should they be captured. (Unschuld 1985:37)

While human sacrifice was abandoned later during the Zhou, this description serves to disabuse us of any idea that demonic healing was a form of psychotherapeutic catharsis. It also reminds us that a spiritual or religious component in healing practices is not necessarily benign or uplifting.

Shamans (*wu*) were often a disruptive presence in Chinese society. Their trance behavior was frightening and they could threaten to evoke retaliation from the supernatural realm without fear of contradiction, since they alone had access. Michel Strickmann (2002:3) suggests that Daoist monks used the excesses of the shamans to present their own version of demonic

healing as a reform. Certainly the use of talismans as a means of communication with the spirit realm was less threatening to the social order than the
unpredictable pronouncements of shamans. Buddhist monks also offered
healing on the basis that illness was the karmic consequences of past life
misdeeds—an etiology just as irrefutable as the spite of demons—and accepted pay for sutra recitation as a means of earning merit for the patient or
deceased relatives. With the Daoist and Buddhist practices, though hardly
scientific, we can see a movement away from the emotional excesses of
shamanic exorcism. Buddhism could make use of the karmic basis of disease as an opportunity for moral instruction, though in practice it must
usually have been rather rote, as well as lucrative for the monk or temple
that would receive the merit-making donations.

Just as Chinese cosmology became progressively more elaborate over
time, so did the demonic realms which became a sort of cosmopolitan
melting pot in which the malign beings of Chinese popular religion min-
gled with those of Daoism and Buddhism, many of the later immigrants
from India, and, later, Tibet. The ghost-queller Zhong Kui was a popular
subject for paintings though the humorous quality of many of these, notably a much reproduced depiction of his sister's wedding in which the
procession is conducted by demons at his behest, suggests that demons
were not always taken entirely seriously (Liu and Brix 1997).

Though demonic healing is not scientific, we may ask if it is spiritual.
The answer depends on the meaning placed on "spiritual." In the sense of
referring to the supernatural, exorcism is spiritual, but because it is mostly
about bitterness and revenge, it is not spiritual in any higher sense. Since it
could not have been efficacious for real diseases, it would have substituted
for other modalities that might have been. One modern apologetic for spirit
healing is that it might have reassured patients who believed in demons.
This romantic idea disregards the fact that exorcism involves stirring up
irrational fears for the benefit of the healer who is paid to take them away.
Another modern rationalization is the suggestion that exorcisms, talismans, and the like might benefit by means of the placebo effect, defined as
healing that takes place when patients are given inert treatments. Though
many misinterpret the placebo effect as the mind somehow being tricked
into healing itself, this interpretation is overly simplistic. In actuality, the
placebo effect is not one phenomenon but several. For example, a disease
may remit spontaneously with or without the treatment. With psychological conditions, such as clinical depression, placebo effects generally occur
but tend to be transient. It can happen that a depressed person's hope that
a treatment will work produces a temporary elevation of mood; without
correction of the underlying neurochemical abnormality, this effect is brief.
Rather than digressing into intricacies of the placebo effect it is sufficient
here to point out that trickery rarely, if ever, results in real healing.

## MEDICINE BASED ON CORRELATIVE COSMOLOGY

We have already warned that common use of the abbreviation "TCM" gives a misleading impression of a unitary system of healing. However most medical theories were ultimately based on the hot–cold, yin-yang, wu xing system, however much they may have differed on details. Accordingly, emphasis in what follows will be on common features.

Interpreting traditional Chinese medicine in modern medical terminology is risky because its concepts of disease are incommensurate with modern scientific ones. Even the supposed functions of the organs in Chinese correlative physiology are entirely different from those established by science. Because there was but limited empirical investigation of anatomy and physiology, the functions of the various organs were based on correlations and are largely incorrect from the viewpoint of scientific medicine. More often than not we cannot determine from the descriptions in Chinese medical texts what disease was being discussed. Hepatitis, for example, was not considered a liver disease, nor one caused by infection. Thus to correlate diagnosis in Chinese texts with modern concepts is often impossible. The diagnostic procedures such as tongue inspection and pulse palpation are entirely subjective. Different physicians would not necessarily come to the same conclusion. That physicians often disagreed on the nature of both ailment and therapy is apparent in case histories. There are however a few diseases with distinctive symptoms, such as smallpox, which we can retrospectively diagnosed with reasonable certainty, especially if they were widespread.

We do not know how the supposed properties of herbs were decided and modern research has so far been unable to confirm the efficacy even of such common botanicals as ginseng and dong gui. It seems likely that the tonic properties of ginseng were originally assumed because of the root's anthropomorphic shape. If this is the case, the prestige of ginseng is based on a correlative association. This is not to say that the voluminous Chinese herbal pharmacopoeias do not contain herbs of therapeutic value. The problem is knowing which ones are worth trying and which ones should be avoided, because there is no evidence for efficacy beyond the author's assertion. Traditional texts rely entirely on authority and anecdote, neither of which is evidence in a scientific sense. As already pointed out, the common argument that Chinese medicine has lasted five thousand years and therefore must be effective is not substantive. First, the earliest sources for Chinese medicine are at most twenty-five hundred years old, not five thousand. Second, once a remedy was recorded and attributed to a cultural hero such as the Yellow Emperor it was unlikely to be reevaluated. Finally, the life expectancy in the late Qing dynasty was less than thirty. Malnutrition, contaminated water, and hookworm infestation were major causes of death. Simple public health measures would have drastically reduced all of these, had their basis been understood. Traditional medicine based on correlative metaphysics offered no insights into such everyday menaces as walking barefoot in fields fertilized with human feces, permitting the larvae to pass through the skin of the feet, into the blood to be coughed up and swallowed, thence to their destination in the intestines where they feasted on the blood of their unfortunate host.

Admitting the limitations of traditional medicine should not discourage us from hoping that much of value to human well-being will eventually be extracted from them. Chinese medicine and Ayurveda are finally receiving scientific study and it is reasonable to expect that some of their remedies will eventually become part of scientific medicine. Progress is likely to be slow because of the high cost of clinical trials, which are extremely labor-intensive. For most who use Chinese medicine in the developed world, the choice is made on the basis of belief, not evidence. For most in China, only the most basic scientific medical treatment is available or affordable.

Like the cosmology it was derived from, the theoretical formulation of Chinese medicine in terms of yin-yang, wu xing, the six forms of qi, and similar correlative categories arose during the late Warring States and the Qin and Han dynasties. Its rise was possible because economic development allowed the medical mainstream to shift from shamans and itinerant medicine peddlers to the literate who could be educated in the complexities of correlative metaphysics. Unschuld (1985:67ff) correlates the rise of these doctrines with Confucian concepts about the proper ordering of the state. Unquestionably, correlative medicine was one of many ideological tools for stabilizing state control.

## DISEASE AND THE ENVIRONMENT IN CHINESE MEDICINE

Wind was from early on considered a significant pathogenetic factor in Chinese medicine. Unschuld (1985:70) regards this as a transitional phase from earlier attribution of disease to demons to later attribution to natural forces. We see here the pervasive assumption that well-being, both personal and social, depends on maintaining accord with natural forces. In practice this probably meant sheltering from winds of certain directions and following sumptuary rules and other prescriptions for seasonal behavior, as already discussed in the chapter on cosmology. Also at this time, notions arose that individual susceptibility determined vulnerability to these environmental and behavioral factors.

While some details of the correlative regulations—such as limiting music—now seem odd, it is striking how closely the underlying assumptions resemble modern ones. Many now believe that environmental influences such as stress, pollution, and improper diet can cause illness and that these can be counteracted by adherence to certain principles of life, often conceived to be natural. These rules, while not openly justified on the basis of asceticism, do prescribe behaviors such as exercise, restriction of calories, and elimination of saturated fat, that require considerable self-restraint. In both cultures, behaviors that maintain health are also to be followed for their spiritual benefit; indeed, physical health and a moral mode of life are not sharply distinguished. Among many Westerners, selection of foods deemed to be natural, such as fruits and vegetables, preferably organic, and animals raised in supposedly cruelty-free settings, is considered part of a healthy and ethically correct lifestyle. Chinese dietary concepts were quite different as we shall see; except for Buddhist monastics there was little inhibition about eating animals at any time in Chinese history. What has not changed is the archetypal belief that living in a natural way will enhance longevity. While there is scientific basis for current ideas of healthy eating and they do extend life on a population basis, the magnitude of this effect is smaller than most assume.

The idea of living in harmony with nature became part of popular Western culture—and marketing—in the sixties but had been advocated much earlier by Rousseau, Thoreau, Emerson, and others of the romantic movement. Though the Western discourse is far more prolix, the resemblance to Daoist philosophy is striking. This resemblance is not entirely fortuitous as Chinese and Indian philosophy were important influences on the romantic movement, which in turn paved the way for Asian philosophy to be taken up in the West.

Though the parallelism in traditional Chinese and contemporary Western thought regarding cause and prevention of disease is striking, we must not ignore the fact that the differences in detail far overshadow the similarities. We no longer regard wind as a cause of illness and what is feared with regard to the environment is not exposure to the elements but the harmful residual of human activity such as pollution and radiation. The extent to which these are actually hazards to most is uncertain but they indicate that the fear of disease from harmful external factors persists after thousands of years.

While now the earth and nature are conceived as inherently peaceful, it was otherwise in the traditional Chinese worldview. The power of nature was frightening but this power could be used for one's benefit under proper circumstances. The shamanistic belief that eating an animal will help one attain its attributes still persists. Thus deer penis is widely sold as a virility enhancer, tiger bones refresh one's spirit, turtle soup gives one the longevity characteristic of that animal, and so on. (Dried deer penis was featured prominently in the Urumchi Airport shop when I was there recently and was quite expensive.) Sometimes the presumed benefits of a food are based on its appearance; walnuts and pig brain are thought to aid intelligence. Such beliefs are correlative in nature and form an important part of what many Chinese regard as healthy lifestyle. Distasteful as deer penis soup may seem to Westerners, its consumption is based on the belief that natural products can positively affect human bodily functioning; the same assumption that underlies use of supplements in the West.

China possessed an extremely rich system of hygiene in which what would now be termed "lifestyle" was linked to health and longevity. This was particularly the concern of Daoist practitioners and was largely separate from the mechanistic systems of acupuncture and herbalism which were called upon to cure disease. Because these hygienic practices emphasized balance and moderation they have some similarity with modern conceptions. No clear distinction was made between food and medicine, anticipating the growing emphasis on nutrition in scientific medicine. Nutritional ideas often had philosophical bases as we see with the Daoist practice of avoiding the "seven grains" and the commitment of East Asian Buddhism to veganism. Although Daoist hygienic practices have spread around the world in the form of tai chi and qi gong, these are greatly simplified versions of the extremely elaborate exercises of Daoist monks.

## CORRELATIVE ANATOMY AND PHYSIOLOGY

The primary pathogenic factors in the medicine of systematic correspondence were six environmental factors whose excess could cause illness. These were yin qi, yang qi, wind, rain, twilight, and daytime brightness (SCC VI:6:43). Common to all is transmission through the air. The character for qi, itself depicting the vapor over cooking rice, indicates the importance of air and breath for the Chinese. This seems to be equally the case in ancient Greece and India where the respective terms *pneuma* and *prana* have a similar meaning (SCC VI:6:43). This emphasis on breath should not surprise us since breathing is the most visible physiological process and the one most obviously essential for life. Indeed, until recent life-support technology, the difference between being alive or dead was breathing. Meditation involves regulation of breathing above all so that breathing has spiritual as well as purely physiological significance. Anyone who has been taught correct meditation technique recognizes that proper breathing is at least as important as the concentrative mental component, though the latter tends to be emphasized in popular Western accounts.

These external factors caused disease through their actions on the body. Excessive yin qi caused cold illness while too much yang qi caused hot illnesses. This hot and cold formulation is found in other cultures and is, I suspect, a very early belief onto which more elaborate correlations were later imposed. The hot–cold system is quite simple to understand and the treatment of cooling or warming follows directly from the etiology, though how a herb comes to be classified as hot or cold is not self-evident. Such simplified etiological models are still widely used in Chinese herbology (Yang 2002).

Bodily dysfunction was explained in reference to the five organs—heart, lungs, kidneys, liver, and spleen and the many meridians or conduits that connect them. Each organ correlated with one of the five phases (wu xing). The courses of these meridians are diagrammatic rather than fully anatomical. Indian medicine has a somewhat similar system of virtual conduits in the body termed *nadi*. These were correlated with the Ganges River and its unusual flow pattern (White 1996:226–229). It seems likely that in both cultures these were inspired by observation of nerves and blood vessels, which were then analogized with rivers, which also have a dendritic geometry. Though we might expect that the river analogy would have led to a clearer idea of the function of the circulatory system, investigation did not proceed beyond the correlative associations. Yet military medical practitioners must have had some practical idea of blood flow.

Chinese medicine regularly refers to the "five organs" as it does to five phases, five musical notes, five flavors, and so on. Assignment of functions to these organs was not based at all on empirical observation. Despite the seemingly anatomical nomenclature, the heart, lungs, liver, spleen are not physical organs but metaphysical constructs and do not correspond in function to the Western organs of the same name. The selection of five organs represents the preference of the Chinese for this number; there are far more organs visible to even casual anatomical study. The system is particularly frustrating to those trained in Western medicine, because it clashes so completely with the actual physiology of these organs.

An example showing how little Chinese notions about these organs accord with Western ones is this didactic verse from the *Su Wen*:

> The heart . . . its fullness manifest itself in the blood vessels . . .
> The lung . . . its fullness manifests itself in the skin . . .
> The kidneys . . . its fullness manifests in the bones . . .
> The liver . . . its fullness manifests in the sinews . . .
> The spleen . . . its fullness manifests in the muscles . . .
> (Unschuld 2003:110)

The association of heart and blood vessels is in accord with modern understanding, not surprising since large blood vessels are connected to it. The kidney does play a role in bone maintenance but this could not have been known because the necessary biochemical discoveries were made in the twentieth century. The liver and spleen have no specific actions on connective tissue or muscle. This is simply an imaginative set of associations, without empirical, or even observational basis. Schemes such as that quoted above, analyze the body in terms of components, just as scientific medicine does; there was nothing particularly spiritual about them. Though the theoretical basis was metaphysical, in practice Chinese medicine was as mechanistic as that of the modern West. If, for example, a condition was diagnosed as being due to a deficiency of kidney yang, acupuncture needles or herbal medications were employed to strengthen this. It is doubtful that it was "holistic" in the currently popular sense of implying that the physician was as concerned about the patient's mental and spiritual condition as his or her physical malady. Indeed, because of the pervasive Chinese proscription of body contact, diagnostic procedures employed such as palpation of the pulse, visualization of the tongue, or inspection of the face (physiognomy) involved minimal body contact. Such procedures as pelvic or rectal examination or even palpation of the abdomen would not have been acceptable because for Confucians even nonsexual body contact was taboo. Thus to claim that Chinese medicine is somehow more whole-body oriented than that of the West is inaccurate. In India, though medicine was less restricted by bodily contact taboos, physical diagnosis was also indirect.

## SOME IMPORTANT MEDICAL TEXTS OF TRADITIONAL CHINA

Although modern treatments of Chinese medicine have proliferated, a true sense of its character requires some degree of familiarity with its foundational texts. Throughout its history, Chinese medicine was largely derived from the principles in these early texts, though many others were published that extended these principles.

The *Huang Di Nei Jing Su Wen* (Yellow Emperor's Middle Classic of Questions and Answers on Medicine) is a detailed compilation of the medicine of systemic correspondences; Unschuld (2003:ix) compares its influence to that of the Hippocratic writings of ancient Greece. A syncretistic work, the earliest sources seem to date from the Latter Han, with some revisions as late as the eight century. Though an early English translation rendered the title as *The Yellow Emperor's Classic of Internal Medicine*, this is misleading. Unschuld's detailed discussion of the difficulty just determining the title of this classic gives a good idea of the problems of translating Chinese medical writings into another language. First, Huang Di was not an actual emperor but a mythical deified sage-king (Unschuld 2003:7–21). Secondly and more importantly, the word translated as "internal," *nei*, seems to have referred to the work as the first section of the book, or possibly as its core. However by the Ming, *nei* was taken to refer to the classic as being about the inside of the body (Unschuld 2003:14–16). Use of the modern term "internal medicine" is misleading as traditional Chinese texts did not divide medicine into the same specialties as does contemporary medicine. While the *Su Wen* contains little that could be considered scientific, it was hardly a casual or random production. The work is immensely detailed and clearly embodies careful observation of illnesses, but the observations are placed within the correlative metaphysics.

Here is a representative example

> The brain, the marrow, the bones, the vessels, the gallbladder, and the uterus,
> these six are generated by the qi of the earth
> . . .
> Hence,
> They store and do not drain.

In contrast, the hollow organs of stomach, large and small intestines and the triple burner and urinary bladder,

> . . . are generated by the qi of heaven
> Their qi resembles heaven.
> Hence,
> they drain and do not store
>   (Unschuld 2003:138f).

Here is the pervasive principle of assignment of objects to heaven or earth. There is an actual observation here in that hollow organs pass along their contents while solid organs such as the liver do not. However, the classification is not consistent because the gall bladder and uterus not only store, they also drain. Even the simple descriptions of the organs' functions are inaccurate because they are forced into a fixed number of each type, then correlated with heaven and earth.

Much of the body of the *Su Wen* is taken up with such assignments of organs and functions to cosmological categories. We could not expect scientific understanding of the body in a traditional culture because the technology for determining microscopic structure and biochemical function did not exist. However, gross anatomy could have developed, yet written descriptions and illustrations of the actual internal organs are generally quite crude. Why this should be so is difficult to establish. It seems unlikely that it was due to religious or ethical inhibition about dissection; fatal war injuries and the grisly methods of execution employed would have been a source for cadavers unprotected by cultural taboos. Most likely it was simply not realized that far more useful knowledge could be obtained by systematic observation than pure ratiocination based on a fixed metaphysics.

Unschuld summarizes this way of thought as follows:

When the concept of a systematic correspondence of all phenomena was applied to the realm of bodily parts and their functions in the second and first century B.C., a common morphological knowledge, unstructured by any theory and listed in . . . vernacular terms . . . had to be categorized to fit the demands of the yin-yang and five-agents doctrine.

. . .

Unlike their Greek counterparts, ancient Chinese authors did not often allow their readers to participate in the processes and arguments that led them to their definitions and interpretations.

. . .

It appears quite impossible to trace the origins of the attribution of the heart, the lung, and other individual core organs with their respective functions. (Unschuld 2003:129f)

These comments get to the heart of the mystery of Chinese medicine: Given the availability of opened bodies for study, why was attribution of function so fanciful? Here is another comment of Unschuld's (2003:130) that can point to a possible explanation:

. . . the ancient Chinese naturalists witnessed a fundamental restructuring of their socioeconomic and sociopolical environment in the time of transition from the Warring States period, with its many units fighting for survival or supremacy, to a well structured, united China where all parts contributed to the well-being of the organism of the state.

In this view, the contribution of the organs to the orderly functioning of the body was analogous to the contribution of individual social units to the empire. Thus, as has been discussed previously, to challenge any aspect of the coherent cosmological system was potentially to threaten the entire basis of Chinese government and society. Referring empirical discoveries back to yin-yang and wu xing was not a choice between alternatives but the only explanatory framework available. The coherence and seeming intellectual power of the cosmology was both a factor in China's stability and a barrier to intellectual innovation.

## ACUPUNCTURE IN CHINESE MEDICINE

The *Ling Shu* or *Spiritual Pivot* (Wu 1993) was the most authoritative classical text on acupuncture, though many others were circulated. It has sometimes been assumed to be the second half of the *Su Wen*, but evidence that the two were associated is at best indirect. The relationship of these texts is complex (Unschuld 2003:5–7, 76–80). The extant versions date from the Song, though the work is clearly a compilation based on earlier sources. In the *Ling Shu*, as in the *Su Wen*, we find a therapeutics entirely based on correlative cosmology, applied to the microcosm of the body. This earth–body correlation is found, for example, in the conception of the acupuncture channels themselves:

There are twelve major channels which externally are in resonance with the twelve major rivers of water, while internally they subordinate the five solid organs and six bowels. . . . The five viscera and six bowels are high or low, large or small, and differ in the amount of valley qi they receive and distribute. (Wu 1993:69)

It is tempting to speculate that the channels were not originally thought of as analogous to rivers but simply as conduits for qi, with the correlation with rivers superimposed on them to fit them into the heaven, man, earth cosmology. There are many such instances in which ideas developed before the Han—yin and yang, for example—became components of much more elaborate systems. However such analogies seem so fundamental in how Chinese expressed their ideas that it is equally plausible that the medical ideas were derived from the analogies. As noted earlier, in India the nadi, or channels for energy flows in the body, were also compared to rivers. Given the similarity between the Indian nadi and the Chinese acupuncture meridians, it has been suggested that the system began in India and traveled to China. The opposite direction of influence is also possible. There is no persuasive evidence for such early transfer so it is just as possible that the systems originated independently. The layout of the channels is different in India and China, but this could be due to separate later development.

Here are examples of diagnostic principles taught in the *Ling Shu*:

Qi Bo said,
"The liver controls and commands. It is the emissary and surveyor of the outside. When one desires to know if the liver is firm and solid, examine to see if the eyes are small or large."
Huang Di said, "Excellent!"
Qi Bo said, "The spleen is the master of the body's protection. It is the emissary of provisions and grain. Inspect the lips and tongue for health and disease and to forecast luck or misfortune."
. . . "the kidneys control the externals. They are the emissary of distant hearing. Inspect the ears for health or disease which may be used to know the kidneys' nature." (Wu 1993:125)

Even setting aside the metaphorical mode of expression—taken from government or military organization—nothing connects these conceptions with the actual function of the organs. In Western medicine the liver is the locus of many metabolic processes that involve production and storage of energy, production of active proteins such as clotting factors, and removal of active substances from the blood. These however bear no evident relation to the metaphor of controlling or commanding. Nor does the liver "survey the outside." Strangest is the notion that the functioning of the liver can be determined by examining the size of the eyes. In fact simple palpation and percussion of the abdomen can delineate the size of the liver noninvasively in most people but even if this had been known, the Chinese taboo against physical contact would have prevented its application. It is true that liver disease may show itself in the eyes in the form of the yellow discoloration of jaundice. But in the *Ling Shu* it is not the color of the eyes but their size. Considering the near universality of hepatitis and endemic schistosomiasis in China, jaundice must have been quite common. Furthermore, when hepatitis results in death there are gross changes in the liver that would have been evident with dissection. Yet these connections were not made. It is hard to escape the conclusion that those who devised the Chinese concepts of diagnosis and treatment simply made them up to accord with the preexisting explanatory system.

The remarks on the spleen might at first appear to anticipate scientific physiology because the spleen is part of the immune system and thus involved in protecting the body from infection. However the remainder of the description of the spleen is not remotely analogous to immune function. Physicians of that era in any country could not have been aware of the nature of immunity, nor the spleen's role in it, which is only apparent with the aid of the microscope. It is important to avoid misinterpreting such statements in early medical texts as somehow anticipating modern knowledge, a temptation even Joseph Needham succumbed to at times.

Here is an example of the sort of therapeutic instructions found in the *Ling Shu*:

"When there are pains in the teeth but no fear of cold drinks, treat the Leg Bright Yang. On the other hand, if there is fear of cold drinks, treat the Arm Bright Yang" (Wu 1993:114).

This is interesting because cold sensitivity associated with a toothache is a sign that the root is exposed; the distinction here regarding response to cold drinks demonstrates systematic observation. When it comes to therapy, however, it is unclear why this would decide choice of the right or left arm for needle insertion and why either would relieve the pain of a decayed tooth.

## FORENSIC MEDICINE

The *Xi Yuan Lu* or *The Washing Away of Wrongs* by Sungzi was composed during the Southern Song and is the earliest extant work on forensic medicine (McKnight 1981). As such, it is of considerable interest. What is particularly striking about this work is its empirical flavor. The *Xi Yuan Lu* does, inevitably, have traces of the medicine of systematic correspondences. For example, the text states that "man has three hundred sixty-five bones, corresponding to the three hundred sixty-five days in a year (McKnight 1981:95). It includes discussion, though brief, of signs that death was caused by a malevolent ghost, or by wind or by other causes not recognized in scientific medicine (Ibid. 139–143). Inevitably, signs of death due to sexual excess are included. In this case, death occurs during intercourse and the penis remains erect after death. On the other hand, if the cause was not excess sexual activity, the penis is flaccid postmortem (Ibid. 151). This seems to be a common myth; a well-known star of Hong Kong martial arts movies who died suddenly, probably from an overdose, was widely rumored to have had an erection when found. This is entirely fallacious; since erection is due to engorgement by blood, the penis cannot be present after cessation of the heartbeat. In this example, a fictitious cause of death—sexual excess—has a fictitious sign.

The greater part of the *Xi Yuan* is clearly based on careful and accurate observation of actual corpses whose death came about by a variety of means. Of decapitation, for example, the text notes that contraction of the severed neck muscles indicates that decapitation was the cause of death while muscle ends even with the skin indicates that the head was removed after death (Ibid. 129). It explains the differences in rope marks when hanging is suicide and when it is murder disguised as suicide (Ibid. 112–114). The section on postmortem examination of women indicates that women attendants should do any procedure that requires touching the body, especially in determination of virginity. Here we find the familiar Confucian prudery circumvented by a socially acceptable means of carrying out an adequate investigation of the death (Ibid. 82–84).

As an example of the careful empirical observation that is found throughout this work, here is an excerpt from the description of bodies killed by tigers:

When tigers bite men, they often bite the head and neck. The upper part of the body will be marked by claws and feet. The wounds are punctures where the bones sometimes show through, on the chest, over the heart, or on the upper legs. There will be signs of the tiger on the ground. (149)

This certainly has the ring of actual observation but the author cannot resist adding a bit of correlative cosmology by asserting that tigers bite on different areas of the body in the beginning, middle, and end of the month. Tigers, needless to say, do not consult calendars. However, these interposed bits of myth do not detract from the suitability of the work as a guide for investigating deaths.

Can we say that this text on forensic medicine has a spiritual aspect? Clearly it is in the Confucian spirit of concern with correction of improper behavior and ordering of the state. In its assumption that strict punishment will follow criminal behavior, it shows the influence of the so-called legalist school. What is striking is that the guilty is to be found, and justice to be served, by careful evaluation of evidence, not by shamanistic visions or by divination. The importance of observation and experience was clearly understood in China, and was commonly applied, though inconsistently.

## DID CHINA DISCOVER THE CIRCULATION OF THE BLOOD?

Needham and his follower Robert Temple asserted that the Chinese discovered the circulation of the blood before the Englishman William Harvey (Temple 1998:123–124). The fallacy in their argument is worth examining here as it illustrates the importance of clearly distinguishing metaphysical and empirical reasoning. Unschuld criticizes Needham's work as:

seeking aspects of Chinese science that seem to anticipate modern science but neglecting those historical thoughts and facts in Chinese medicine that are irreconcilable with what the authors consider to be scientific, protoscientific, or at least rational. (Unschuld 1985:2)

It might be added that this is not unlike the "figurism" of some of the early Christian missionaries in China who sought to interpret aspects of Chinese culture as if they foreshadowed the Christian revelation. Needless to say, interpreting Chinese philosophy and religion as if they were incomplete forms of Christianity did not produce an accurate vision of China. What it did do, as did the scientific figurism of Needham, is encourage sympathy with much of Chinese culture that would otherwise be unappealing to Westerners. While this wish to show the greatness of Chinese civilization is commendable, it would be better served by scholarly accuracy in place of apologetic.

Here are some of the passages Needham translates and quotes in support of his argument that the Chinese knew of the circulation of the blood:

The *yang-ch'i* runs within the blood vessels ... The yang-ch'i circulates endlessly, never coming to a stop except in death.

The heart presides ... over the circulation of the blood and juices, and the paths which they travel.

The heart rules over the circulation of the blood, and the pulse . . . it sends the blood to all the parts of the body . . .
(Needham 1981:100f)

Of this last, Needham makes much of the publication date preceding by several years Harvey's report of his discovery, but since any influence is out of the question, this matter is of no importance. Needham also mentions Harvey's cosmological ideas, which were typical of his time (Needham 1981:103f). To hold metaphysical beliefs does not necessarily prevent empirical observation; Harvey's cosmological beliefs do not vitiate the brilliance of his discovery.

Here is another passage in Needham's translation:

. . . if there is no working of the bellows . . . [lit. drumming, ku] then the blood will not flow round. (Unschuld 1985:371f n.40)

Unschuld points out the distortions in Needham's translation of this passage. First, Needham translates a Chinese word that refers to the pulse, *ku*, as "the action of a bellows" a phrase which seems to embody understanding of the heart as a pump but which is not actually present in the Chinese text cited by Needham. The literal meaning of *ku* is drum or drumming, not the action of a bellows or pump. This analogy is straightforward as the heart does feel like a drum to a hand placed on the chest. Furthermore, the supposed flowing of the blood is, in context, simply one instance of the flows of various sorts of qi, not of actual biological fluids. Thus Needham bends the evidence in a way that will be persuasive to that majority of his readers who cannot access the Chinese text itself.

There is another problem with Needham's hypothesis: even if some Chinese physicians did conceive of the circulation of the blood, nothing further followed from it. A good guess has no usefulness unless it stimulates further work. Harvey's discovery was among the most fundamental ever made about the functioning of the body; without it scientific medicine and surgery would not be possible. In fairness to the Chinese, however, it must be noted that it took the West several centuries to attend to the medical implications of Harvey's discovery.

Chinese medicine did assume that something did circulate in the body. However, what circulated was qi, invisible and undetectable by any objective means. Given that the primary physiological parameter could not be seen, there was no theoretical reason to study the body directly. However, with the example of forensic examination of corpses, we see that the governmental interest in detecting murder overrode metaphysical theory.

Qi is also a concept in Chinese meditation practices and in martial arts. Experienced meditators can evoke a tingling sensation and may also feel a subjective sense of something flowing within the body; these are considered to be the direct experiencing of qi. There is no reason to doubt that these are real as subjective sensations. Such concepts are useful to meditators in developing their practice and so, in this context, qi can be real. However to take this farther and regard these subjective sensations as actual forms of energy that are involved in the mechanisms of disease is without any empirical basis.

## EMPIRICAL MEDICINE IN CHINA: SMALLPOX VACCINATION

Smallpox vaccination originated in China and provides a most interesting example of the interplay of metaphysical and empirical factors in Chinese medicine. The subject has been well studied by modern scholars, though questions remain, particularly as to when and how the practice originated in China. Since smallpox can be visually diagnosed we can be reasonably sure that the traditional Chinese sources are referring to the same disease we term variola (smallpox). It may be that there was not an entirely clear distinction between smallpox and other childhood exanthems (viral diseases characterized by a rash) such as rubella (measles) and varicella (chicken pox). Nonetheless, since the latter diseases are usually nonfatal, smallpox was what stimulated the most medical concern. Furthermore, we can recognize exanthems from descriptions while such diagnostic categories as cold excess and yin feebleness have no correlates at all in scientific medicine. Thus with smallpox we can quite directly consider traditional Chinese theory and practice in comparison to our own.

Immunization has a simple rationale derived from the fact that those who recover from certain viral illnesses never become reinfected. This observation led to development of effective immunization before its scientific basis became understood. It is now known that infectious illnesses generally stimulate a response from the immune system, which can eventually eliminate the infectious agent. Antibodies, which are protein molecules, combine with viruses or bacteria so as to neutralize or kill them. Because antibodies are formed only as a response to infection they are not present unless there has been prior exposure to the disease-causing organism. It is the persistence of antibodies after recovery from certain kinds of infection that prevents reinfection. However because initial antibody formation may take a week or more, death may occur before the body's immune response develops. (There is much more to immunology than this. For some diseases such as HIV and many bacterial infections, the immune response is inadequate to eradicate the infectious agent. With others, such as chicken pox, the virus may live in the host for many decades and later reemerge as the painful rash termed "shingles.") Inoculation with the actual smallpox virus is termed variolation; the term "vaccination" technically refers to immunization with a modified virus but has come in general usage to refer to any sort of immunization.

With smallpox, those who survive the disease never develop it a second time. Thus it made sense to infect healthy people with a mild form, so as to induce permanent protection against the full-blown disease. The early use of immunization was based on this empirical observation but did not require knowledge of how the process actually works to be put to clinical use. The essential technology is a method for weakening the virus so that administering it does not result in severe disease but an antibody response is still evoked. This remains the central challenge of vaccine development today.

Textual references to smallpox in China appear sometime between the third and sixth centuries C.E. [Volkmar 2004; SCC VI:6:117.] and become common after that. Presumably the disease was brought to China over trade routes. Immunization seems to have become widespread in the later Ming and perhaps nearly universal during the Qing. It was officially endorsed by the Kangxi emperor (reigned 1661–1722) whose father had died of smallpox. In an early instance of government concern for public health he stated, "The courage which I summed up to insist on its practice has saved the lives and health of millions . . . " (SCC VI:6:140). One cannot help but notice that the emperor reserved credit for himself, not the inventors and practitioners of vaccination. A theory that variolation was known much earlier in esoteric Chinese religious or Daoist sects but kept secret, has been reviewed by Needham (SCC VI:158) and is not taken seriously by scholars today.

In both China and the West, the actual practical of vaccination developed without, or in spite of, a theoretical basis. Volkmar (91–94) provides a useful compilation of Chinese theories of the pathogenesis of the disease. One theory was embryonic or womb poison, the notion that toxic influences pass from the mother into the fetus and later cause illness in the child. One source of this toxin, though not the only one, was thought to be excessive sexual pleasure, fear of which was always prominent in Chinese hygienic thought. This theory is an example of the assumption, universal until fairly recently, that any disorder in a child is the fault of the mother. The second major theory was "seasonal qi," that is, hot and cold injury. The arbitrary nature of assumptions about hot and cold as causes of disease is shown by the fact that one school considered smallpox a cold illness to be treated by warming while another took it to be a warm illness to be treated by cooling. The third major theory was "five periods and six qi," which assumed that interactions of cosmic cycles accounted for changes in incidence of the disease. These theories could be meshed with each other by making the womb poison hot in character and attributing the time of appearance of the eruption to cosmic cycles. The development of variolation did not seem to cause such theories to be questioned. Indeed, selection of favorable dates was assumed to be as critical for immunization as it was for all other important activities.

Needham gives thorough coverage of the techniques used in China for attenuation of the smallpox virus. The Chinese method was fundamentally different than that discovered by Jenner in England in 1796 because the Chinese used the actual variola virus while Jenner used bovine lymphatic fluid containing the virus of cowpox. This was a safer method because cowpox induced cross-immunity to the variola virus but did not cause serious illness in humans. The cowpox virus was artificially cultivated and over time became a unique form termed vaccinia, thought to be a hybrid of variola and cowpox virus (SCC VI:6:116).

In China, inoculation was most commonly performed by introducing a scab from a smallpox patient into the nose of the person to be immunized. There was considerable discussion among physicians as to what sort of scabs were optimal. Dried ones seemed to be preferred, which would have likely held a smaller inoculum of viable variola virus. It was also recommended that the practitioner carry the material in one's pocket for a time. It was observed that cold exposure prolonged the potency of the scabs while heat shortened it. This was explained on the basis of retention or loss of qi but is surely correct since refrigeration tends to preserve viruses while heat denatures them (SCC VI:6:143). Carrying the scab was undoubtedly an effective means of reducing the virulence of the virus and indicates an experimental spirit on the part of the practitioners of vaccination, even though the mechanism of the procedure was not correctly understood. This provides yet another example of the importance of systematic observation over theory in medicine. Even in modern medicine, we do not understand the mechanism of action of many well-established treatments.

An intriguing variant was the use of cattle lice that were roasted and ground, then ingested orally. It is possible that cowpox viruses in the louse could have served to produce immunity but the roasting might well have denatured them. Nor is it clear that oral ingestion would have been effective. This may simply be another example of the fondness for insects as remedies in traditional China.

## SCIENCE AND SPIRITUALITY IN SMALLPOX VACCINATION

While Chinese smallpox immunization can be described in terms of its scientific aspects, religious belief and ritual surrounded its use (SCC VI:6:156–164). Vaccination was associated with Daoism and its use was accompanied by sometimes elaborate religious activity. Here is an excerpt from a set of such instructions:

For offerings the flesh of two animals made into five dishes, as well as five different sorts of fruit, are required. The altar must be . . . hung and draped with red cloth. The names of the gods are to be written at the top of a piece of red paper . . . (SCC VI:6:161)

The emphasis on the number five and color red indicates that the empirical efficacy of vaccination did not diminish the concern for proper ritual or magical observance. Given the fear engendered by this terrible disease that disfigured many of those it did not kill, it is not surprising that deities associated with smallpox were well represented in the Chinese pantheon. Smallpox deities were multiple. One, Toushen Niang-niang, translated by Needham as "the Comptroller of Smallpox" had four sons, each representing a different form of exanthem. There were even temples devoted to a historical, later deified, "immortal teacher of inoculation" (SCC VI:6:156ff). Talismans and incantations against smallpox were published and no doubt widely used. Pages from two such manuals are reproduced in Needham (Ibid. 159f. Figs. 9, 10). They are fairly typical of such productions in combining normal characters in special ways that would have seemed magical to many Chinese readers. For example, several large graphs combine the character for ghost, *gui*, with evocative phrases such as "the pustules come out and it is not pleasant," "the scabs fall off," and more obscure ones such as "transforming steam."

Thus we find three sorts of themes with regard to smallpox: an effective practical technique, a group of theories of pathogenesis one of which, the womb poison theory, was partly mechanistic though incorrect, and ethico-religious theories that attributed the disease to sexual or other transgression. In general, Chinese did not feel any necessity to choose between what seem to us to be inconsistent theories. In this way science and religion coexisted rather than competed.

We have seen that China did not lack skeptics regarding the role of the supernatural in human affairs. Needham quotes a 1741 text by the physician Chang Yen who criticizes physicians who do not inform themselves of the new techniques of inoculation but simply rely on old books:

Not only are they unable to restore the health of their dying patients, they blame their failures on mischief wrought by heavenly visitations. (SCC VI:6:136)

Here we notice awareness of the importance of new methods, recognition that classical texts have their limitations, and awareness that astrology provides a convenient pretext for blaming one's own failings on matters beyond one's control. Practical observation and openness to new methods are always available to the human mind and emerge recurrently, even if without theoretical justification. This sort of prepositivist empiricism tends to arise when the cause and the outcome are close together in time and the connection readily observable. We shall see that this was also the case with ceramic technology.

The Chinese did have a theoretical basis for inoculation, though it was not substantive. Despite the elaborate religious hocus-pocus surrounding smallpox vaccination, intelligent physicians were able to make critical observations and apply them to prevention of this terrible disease. Moralistic theories did not prevent recognition of the two key facts: That nearly all children would develop smallpox and that survivors of mild cases of smallpox were protected from death from this disease. Indeed the effectiveness of variolation could be explained without the need to refute the womb-poison theory, which held that the rash represented expulsion of the poison transmitted from the mother's womb. Some attributed the persistence of the toxin to failure to remove meconium or blood from the mouth of the fetus (SCC IV:6:130). (That rashes, such as acne, are caused by internal toxins or improper nutrition is still widely believed in the West, though scientifically refuted.) The effectiveness of inoculation was brought into accord with the womb-poison theory by interpreting the milder skin eruption as providing a means for the toxin to escape with less systemic effect. Though incorrect, the womb-poison theory was in part mechanistic.

While it is unclear whether the womb-poison theory stimulated practical techniques of variolation, it may have been an effective argument for proponents of the immunization to use it against their conservative opponents. One source of this opposition was religious as immunization contradicted the teaching of both Daoism and Buddhism that disease was the result of transgression in this or previous lives. Since expiation typically involved purchase of redemptive services from the religious institution, the self-interest is obvious. Yet this idea of disease as punishment is not unusual as a folk-belief today.

The use of variolation shows that medicine in China was able to produce benefits in public health and that the emperor approved this. John Arbuthnot in 1722 studied London death records and calculated that the mortality rate of smallpox was one in nine, but that of inoculation far lower. The same would likely have been true for China (SCC VI:6:147). In the sense of benefiting millions, variolation was a spiritual procedure, more so than religious rituals or talismans. Ultimately, the spiritual value of medicine is determined by its human benefits, not by its relation to religion or philosophy. This is not to say that the latter have no value in medicine; they clearly do. However they do not substitute for effective scientifically validated treatment.

## TRADITIONAL INDIAN MEDICINE

Like traditional Chinese medicine, that of India is enjoying a revival in both its native country and in the affluent West. The Sanskrit term for traditional medicine is *Ayurveda*, commonly translated as "science of life." Indian medicine is first attested in the *Atharvaveda*, a miscellaneous collection that includes spells, mantras, and herbal prescriptions. Medicine became formalized about 500 B.C.E.; texts began to multiply beginning in the first century C.E. (Desai 1997:669). Hinduism attributes the invention of medicine to the mythical *rishis*, ancient religious sages, comparable to the ancient Chinese cultural heroes but more explicitly religious. Much of the early development of Indian medicine however seems to have occurred in Buddhist monasteries, or at least that is what the written record suggests. The attribution to rishis, like the Chinese attribution to cultural heroes such as Huang Di (the Yellow Lord), protects medical knowledge by giving it spiritual sanction but at the same time by lending cultural authority to the earliest forms of medicine tends to inhibit progress.

The earliest Hindu texts, the *Vedas*, particularly the *Atharvaveda*, contain much material intended for healing, consisting of hymns and mantras, the latter being particularly prominent in Indian religion since the earliest times. Zysk characterizes Vedic medicine as magico-religious:

Causes of diseases are not attributed to physiological functions, but rather to external beings or forces of a demonic nature who enter the body of their victim and produce sickness The removal of such malevolent entities usually involved an elaborate ritual . . . nearly always necessitating spiritually potent and efficacious words, actions and devices. (Zysk 1993:8f)

Symptoms were correlated with possession by specific disease demons. Zysk contrasts this magico-religious medicine to empirical-rational medicine that was based on observation and experience. It is this latter that constituted Ayurveda. However, the rational or purely metaphysical component predominated over the empirical. Perceptive observations were made but they tended to be fitted into the preexisting correlative system. The separation of rational from supernatural healing modalities was a more manifest process in India than in China, simply because the Indian canonical texts contained far more religious elements than those of China. Yet the Ayurvedic texts remained metaphysical. Nonetheless, in the gradual predominance of rational medicine over that based on the supernatural we can see that in India medicine gradually separated itself from purely religious principles just as it did in China.

## DIAGNOSIS AND TREATMENT IN AYURVEDA

Ayurvedic pathophysiology explains disease on the basis of *doshas* or humors. These are: *vata* (wind or movement), *pitta* (bile or heat), and *kapha* (phlegm or body substances) (Desai 1997:670). The similarity of the doshas to the Greek concept of humors may be due to transmission between Greece and India via Persia (Iran). This theory of pathophysiology is similar to that of China in that disordered inner substances or forces are assumed to be the cause of disease. Frequently the internal imbalance is induced by harmful external influences. Both systems developed these notions in elaborate detail. While yin-yang and wu xing do not have exact analogies in Indian metaphysics, the underlying way of thought is similar.

Health in Ayurveda is a state of harmony or balance of the doshas while disease is an imbalance. These imbalances can be caused by a variety of factors including heat and cold, intense emotion, improper diet, colors, and scents. Though all are exposed to such adverse factors, the nature of the disease produced is determined by that individual's pattern of doshas. All people are classified into types based on which dosha is predominant. One's dosha determines what lifestyle choices each individual should make including foods, yoga postures, colors, and scents. (Lad 1984:144–150). As examples of color effects, red alleviates vata and kapha, but can increase pitta. Gemstones play an important role in both healing and in neutralizing unfavorable factors in one's horoscope. With respect to doshas, red coral calms pitta and helps control anger and jealousy. Ruby is good for vata and kapha but bad for pitta. Agate soothes fear in children and stimulates spiritual awakening. These effects tend toward the psychological because they are from a contemporary manual of Ayurveda that emphasizes aspects attractive to Westerners. This focus on the minutiae of daily life and how they affect well-being has, however, been prominent in Ayurveda in the past. In light of scientific medicine, it is more plausible that such measures might have psychological benefits than that they would cure physical diseases. The emphasis on calming is particularly appealing. In its future in the West, Ayurveda may have more application to restoration of well-being than therapy of specific diseases.

In one important respect the Indian approach to lifestyle is quite different from that of the contemporary West—personal choice is limited. Thus someone with a kapha constitution should eat raisins and apples, but avoid bananas, avocado, and oranges. This system is correlative, detailed lists indicate which foods were permitted or not permitted for each dosha. Although it is obviously not derived from modern nutritional research, Indian food with its vegetarian tradition is far closer to contemporary notions of healthy diet than Chinese cuisine with its emphasis on pork and lard.

As currently practiced, an Ayurvedic treatment plan consists of an initial phase of purification, termed *panchakarma,* which may involve fasting, induction of vomiting, enemas, and purgation. Patients undergoing panchakarma, not surprisingly, may feel worse before they feel better. This is interpreted as toxins being released. Such practices are found in Western prescientific medicine as well. The belief that disease involves accumulation of toxins or impurities is universal. From the view of modern medicine, in someone truly ill these therapies may make matters worse by inducing dehydration and electrolyte depletion. However panchakarma may have served a function like the spas of nineteenth-century Europe in which the wealthy and self-indulgent, feeling out-of-sorts after months of high fat meals washed down with copious quantities of wine, and followed by cigars, were placed on large amounts of water and bland food.

Following panchakarma, a phase of treatment with medicinal herbs and/or minerals begins. Unfortunately, in India as well as China, mercury was regarded as a near-universal remedy. It has been speculated that the bright red color of cinnabar, a compound of mercury, might have seemed to resemble blood and thus be health promoting. How often people in these societies ended up with some degree of mercury poisoning is yet to be investigated. Ayurvedic physicians still hold mercury in high esteem, claiming contrary to the principles of chemistry that it can be rendered nontoxic. The example of mercury use shows the potential hazard of relying on correlative associations. That cinnabar resembles blood has no relation to the effects of ingesting it.

Though it shared the metaphysics based on three that was pervasive in India (three gunas, three doshas, the three principal gods, Brahma, Vishnu, and Shiva, and so on), Ayurveda was not itself religious in practice. Rather, Ayurveda became a set of diagnoses and therapies that were essentially technical. It is true that diagnoses often were related to the perceived spiritual and psychological state of the patient. Then as now stress was considered a cause of disease. Religious notions of disease continued and were not supplanted by Ayurveda. As noted previously, religious diagnoses were not always sympathetic to the patient as any disease might be attributed to unfavorable effects of past life transgressions. Karma is a sophisticated concept philosophically but in practice it tends to degenerate into an excuse for blaming people's misfortunes, including disease, poverty, and low social status on their own past life actions.

## SURGERY IN CHINA AND INDIA

In China, surgery fell into disrepute as a result of a possibly mythical episode when a king died following an operation. However, the restricted use of surgery is consistent with the strong Chinese taboo against touching or altering the body. This prohibition is found in the beginning of the *Filial Piety Classic (Xiao Jing)*:

Our bodies, skin, and hair are given to us by our parents; we dare not injure or wound them. Here filial piety begins.

Of necessity, adherence to this principle was often symbolic, such as abstaining from trimming the nail of the fifth finger, upon which the wealthy might place a protective case of precious material. While such practices may have had spiritual value in reminding one of one's parents, the taboo against bodily therapies greatly inhibited development of both physical diagnosis and surgery in China. Scientific medicine, in contrast, is highly invasive, sometimes needlessly so. Yet without this willingness to search out disease wherever it lies hidden, modern medicine could not exist. The recent trend is toward developing less invasive diagnostic and surgical methods, though the necessary technology is significantly more expensive.

Daoism taught complex meditation practices that were claimed to modify the internal organs. Such beliefs indicate that China never recognized a clear distinction between anatomy, which can only be altered by physical means, and physiology, which might be affected by drugs or acupuncture. Organs were not so much physical structures as the loci of various forms of qi. This is a form of mind-body holism, but one that could not distinguish those conditions that might be healed by mental change, such as anxiety, and those that could not, such as infection or cancer. Unavoidably there was some form of surgery in China, particularly in the military. A story about China's surgeon/culture hero, Hua Tuo, relates him removing an arrow from the arm of Guangong, the hero of the *Romance of Three Kingdoms (San Guo Yan Yi)*. Guangong was so calm that he played chess throughout the procedure.

Not surprisingly, knowledge of anatomy developed to a greater degree in India than in China. The initial reason for this was not scientific curiosity but religious, but in a way that now seems distasteful. Anatomy was learned by separation of bodies of both horses and humans during religious sacrifice. The rituals required that each part be named as it was cut (Zysk 1993:7). This seems to have been the origin of Indian anatomical knowledge, though it can be assumed that a practical understanding of anatomy was attained by battlefield surgeons who would have put it to use in their treatment of the wounded.

In India surgery became highly developed, although here too, some limits were set by inhibitions regarding bodily contact. One of the major Ayurvedic works that is attributed to Susruta (Dash 1997:927) provides a particularly extensive discussion of surgery, though characteristic of correlative works, the creation of the universe is discussed in the same section as human embryology. Most significantly from our perspective, Susruta held that dissection of dead bodies was essential to the training of surgeons. He also described surgical procedures and appropriate instruments. Thus we can see another example where the importance of empirical methodology was recognized but without casting doubt upon correlative medical theory. As with forensic pathology, which we have already discussed in relation to China, surgery lends itself to empirical methodology because valuable data can be obtained by careful, but unaided observation. For many other conditions such as infections or hormonal disorders, understanding could not be achieved until microscopy and biochemistry had developed.

Ayurvedic parishioners insist that their work is based upon spiritual principles. However much that is spiritual in Ayurveda is derived from other aspects of Indian spiritual tradition, particularly meditation. Certainly Ayurveda is compatible with meditation and was routinely combined with use of mantras and amulets. Modern versions of Ayurveda emphasize these spiritual practices because they are a major part of its attraction for people outside India. Those who are ill want to be cured, of course, but also want satisfying explanations for why they are sick and what may lie ahead for them. The metaphysical pathophysiology based on doshas seems to appeal to some who find scientific explanations unsatisfying.

A basic function of spirituality is to provide solace in frightening life situations. Disease always raises the possibility of death and was particularly frightening in the past when effective treatments were few and prognoses unreliable. This prompts us to ask whether traditional medicine somehow was better at easing the patient's fears than scientific medicine. On the one hand, explaining a disease within a belief system based on micro- and macrocosmic parallelism may have made it seem a part of the natural pattern rather than an unaccountable violation of it. On the other hand, we can wonder if such abstract philosophical explanations made much difference to those facing imminent death or experiencing intractable pain. With respect to exorcism, it is common in the West to sentimentalize such rituals but as we have seen, shamanic rituals must have been fearsome events, far more stressful than sitting in a therapist's office with a cup of herb tea and a box of tissues nearby. Beyond this, we know little of the psychological effects of such rituals.

As to the efficacy of Ayurveda, we have little usable data. Since the Indian terminology of disease, like that of China, cannot be translated into the nomenclature of scientific medicine we usually cannot determine what disease was being treated. If a patient was treated and suffered permanent disability or death, we cannot tell whether the disease was truly incurable or whether the treatment hastened the patient's exit. Nor would the practitioner know the effects of treatment, given the lack of the systematic data necessary to assess prognosis. No doubt experienced physicians were able to judge disease severity, though with limited accuracy. Failure to cure would have been blamed on the disease rather than on the treatment, particularly when the physician's own life might be at stake with a high-ranking patient. In both India and China, one occasionally finds ethical admonitions against physicians accepting care of patients with incurable diseases. This now seems harsh but protected the patient's family against futile expenditure and the physician from blame. The argument that Ayurveda developed over five thousand years is no more valid than the same argument for Chinese medicine.

As with traditional Chinese medicine, the ultimate place of Ayurveda in modern heath care is unclear. It is plausible that it will have useful remedies to offer, though separating what is truly effective and safe will be an extremely slow process because of the high cost of such research. Though Ayurveda has not caught the Western imagination to the same degree that Chinese medicine has, it merits equal consideration.

## TRADITIONAL MEDICINE IN THE MODERN WORLD

As we have seen, the Chinese medical system was not monolithic, nor were all physicians oblivious to its limitations. Some were acute observers who must have realized that following the textually sanctioned regimens often failed to produce a happy outcome for the patient. Unschuld (1985:197) describes vividly the dilemma posed by growing awareness of the inadequacies of the medical system:

If the first half of the history of the medicine of systematic correspondence, from the Han to the Sung, is characterized by a naïve application of the theories . . . the second millennium following the Sung period reveals a steadily growing unrest, reflected in various reductionist etiologies [such as hot and cold], in debates about anatomical and physiological details and in attempts to expand the theoretical system. The diversity of schools and their conflicting views during the Ming and Ch'ing periods convey the impression that the conceptual framework of systematic correspondences at this time was nothing more than a complex labyrinth, in which those thinkers seeking to solutions to medical questions wandered aimlessly . . . unable to find a way out. Such a solution came only with the collapse of the Confucian social order . . . that had prevailed for centuries.

Thoughtful physicians felt that their system was under strain and proposed variant theories, but they were unable to move past the orthodox Confucian cosmology. From a modern scientific perspective, Chinese medicine moved sideways, adding to its doctrines while never getting closer to a consistent empiricism. For this reason, despite the considerable intelligence applied to medical questions, it was not possible to make any real progress with them. Western medicine was accepted into China without extreme resistance, no doubt because intellectuals had become frustrated by the inadequacies of the traditional system.

The story does not end here however, because most Chinese continue to believe in their traditional medicine. One often hears Chinese remark that their medicine does work but more slowly than Western medicine, that it is good for chronic diseases, and that it has no side effects. I have heard these even from the daughter of a Western-style physician who occupied a prominent position in the Beijing medical hierarchy. The tendency of Chinese culture to add rather than replace is apparent here because many Chinese avail themselves of both Chinese and Western medicine. Few physicians practice both because each system requires many years to learn and because to practice two systems whose theories are entirely incompatible is an almost impossible mental feat. Some physicians trained in Western medicine do dabble in Chinese medicine but they utilize isolated bits—a few herbs or a few acupuncture points—not the entire system. The reverse is not true. Traditionally trained healers cannot practice Western medicine.

Regrettably, what is sold as Chinese herbal medicine is often adulterated with pharmacologically active Western drugs, such as an aspirin analogue or estrogen. Thus what is presented as traditional medicine can surreptitiously make use of Western therapies. This situation is rather dangerous as purity and dosage are uncontrolled and the person does not know what he or she is getting.

Why traditional medicine continues to thrive in China and elsewhere is a phenomenon of great interest that has not been adequately explored. Here we can offer only a few conjectures. One obvious motivation is cultural affirmation. That their millennia-old traditional medicine may still have a place in modern healing is a source of pride to Chinese. Also, using this medicine is a way of being Chinese, like using chopsticks or celebrating the lunar new year. All cultures have activities that provide a sense of belonging; for Americans these include eating turkey on Thanksgiving and cheering at baseball games. Use of traditional medicine is comfortable for many Chinese, because it is familiar and because it fits the metaphysical system that underlies much of the culture on both conscious and unconscious levels. An additional obvious factor behind promotion of Chinese medicine is the economic interest of those who make their living by it. None of these factors have any relation to actual efficacy and safety. The challenge for Chinese traditional medicine in the twenty-first century will be to find a way to preserve its cultural value while subjecting it to scientific testing.

In the West, motives for use of Chinese medicine are somewhat different. To the extent that cultural affirmation is involved, it is to affirm one's openness to a culture different from one's own. There is also the assumption that herbs or acupuncture are somehow more "natural" than Western surgery or pharmaceuticals. The appeal of trying something that is new and exotic is likely an important motivator, though not necessarily acknowledged. The experimentation with drugs and sexuality that became widespread during the sixties was an expression of a greater willingness to experiment on one's own body. The popularity of spas and the fitness movement can be traced to the same culture change.

The widespread belief that Chinese medicine is somehow more "natural" than that of the West is based more on its being ancient and exotic than upon close familiarity. Whether such treatments as mercury, dog urine, dead insects, or snake blood are "natural" is dubious. These clearly do not accord with the Western preference for plant-source supplements and for this reason, tend to be suppressed when Chinese medicine is offered to non-Chinese. However the idea that Chinese medicine is natural can persist because there are no clear-cut criteria for applying this term, which is basically in the eye of the beholder. At root the desire to use natural remedies comes not from empirical evidence but from the same desire to live in harmony with natural laws that we find in Laozi's *Dao De Jing*, more widely read now than ever before in history, and throughout Chinese philosophy. The yearning to find a way of life congruent with the underlying pattern of the universe is perennial, but what is considered a natural life changes over time. This is not to disparage the desire to live naturally but to acknowledge the fluidity of means for doing so.

Spiritual and scientific values do not always coincide. That a treatment is "natural" does not ensure that it is safe or effective. Vitamins are thought to be natural and healthy but recent research has demonstrated that certain vitamins make lung and prostate cancer worse. Too much vitamin A makes bones fragile; the list could go on. This issue is at the heart of the distinction between spirituality and science. Many want to live in accord with spiritual principles and thus face discomfort when important beliefs—such as the health value of vitamins—are not confirmed by science. This is not unlike the dilemma of traditional Chinese physicians as they recognized that correlatively based therapeutics often failed.

Some spiritual practices are beyond refutation. If a person finds prayer or meditation enhances his or her life, no further proof is pertinent. However to advocate visualizations as able to cure cancer in place of chemotherapy is a quite different matter because of the potential harm.

Many now feel that scientific medicine has become detached from any spiritual ground. I do not think that physicians and others in the system are less spiritual than any other group. It is true, however, that medicine as a body of knowledge does not have spiritual content. Rather, spirituality in connection with disease and healing must be brought in from outside of science. We have seen that this was the case with Chinese medicine with respect to smallpox. While many of the practices now seem superstitious, the scientific advance of immunization did not require elimination of religious practices that the people wanted. What is different now is that each must find his or her own way to integrate spiritual life with science. Though some scientists have argued that humanity should simply abandon religious ideas, this is not really possible. All of us have values, conceptions of how we ought to live, even though these may seem secular, they are still spiritual.

The coherence of traditional correlative thought appeals to many in the modern world. This may be the wish to recover what Freud called the "oceanic feeling," the vicarious experience of primal unity. Freud was disdainful of this desire, which he considered infantile. Yet the current revival of interest in mysticism shows that Freud was too quick to dismiss this longing. Much of the appeal of Chinese and Indian spirituality is that they offer both philosophical doctrines—the Dao and the Atman—together with techniques for attaining this feeling of oneness.

This revival of interest in attunement to an all-encompassing spiritual principle seems to me an inevitable response to the strict materialism often attributed to science, or capitalism, or both. To apply Chinese cosmology, the yang of materialism is now waning and the yin of mysticism is waxing. However, this surge of interest in correlative thought is possible because the restrictions it imposed on earlier life have been forgotten. Coherence is satisfying until one comes up against the limits it places on imagination and behavior.

# Chapter 9

## Chinese Ceramic Technology

In many areas Asia was ahead of the West until science and technology began to flower in the eighteenth century. Ceramic technology provides an instructive example. Until the eighteenth century Chinese ceramics were superior in technical and aesthetic quality to all others. Europe only caught up by studying and emulating Chinese production methods beginning in the mid-eighteenth century. The most common English term for porcelain, "China" is due to the association of this technology with that country. The relevant question here is the opposite of that of Joseph Needham, who sought the factors because of which Chinese science never developed to its full potential. With ceramics, what is to be explained is why the technology was superior to any other in the world.

While there is a tendency to regard technology as a lower order of knowledge than science, from the point of view of empirical scientific method, this assumption is unwarranted. Because technology requires testing methods to see if they give the result intended, it is inescapably empirical. Indeed, it is in technology that we often find the clearest scientific reasoning. Cosmological speculations are limitless because their concepts can neither be confirmed nor refuted. With technology, a method either succeeds or must be abandoned. In the example chosen for this chapter, ceramic manufacture, a vessel emerges from firing intact, or it droops or shatters. Patrons and customers will not tolerate failed results. Medical failures can usually be explained away as lack of compliance or incurability, erroneous divination can be blamed on the interpretation rather than the method, but in technology bad results are too obvious to explain away and might be harshly punished in the ancient world whose governments felt no concern for human rights.

The production of ceramics was widespread in the ancient world. In China, pottery production began at least eleven thousand years ago and was progressively refined. An idea of the extent of ceramic technology in China is shown by the massive length of the volume on this subject in the *Science and Civilization in China* series—eight hundred pages (SCC 2004). Technology seems to grow at an exponential rate, rather than a linear one. Hence the earliest improvements took literally thousands of years. From the beginning there were a variety of styles and manufacture methods, though these grew progressively more numerous. Only in the nineteenth century did Chinese inventiveness in ceramics begin to wane, partly due to the technology reaching its highest possible degree of refinement and possibly also because of the general decline in government effectiveness.

There are several stages in the production of a ceramic. These are described in detail in (SCC V:12:1–86 and passim). First is selection of the clay or clay-like material, which is essentially rock that has been ground down into a fine powder by a variety of natural processes, including abrasion by water or ice, alternate freezing and thawing of water in rock crevices, and chemical degradation by water with an acid or alkaline pH or high salinity (SCC V:12 45). In China, though not always in the West, pottery production tended to be located in areas with large deposits of suitable clay. Usually each region had specific types and styles set by the possibilities of its clays. The most prominent pottery-producing region in later Chinese history was Jing De Zhen; the delicacy, luminous white, and elegant glazed decoration from this kiln has never been equaled. In the nineteenth century, Europeans began to visit Jing De Zhen to learn its technology; it is from this that Europe learned to produce the very thin, translucent but strong porcelain treasured by the wealthy of Europe and America.

The important characteristics of clay are its plasticity, that is, its ability to be formed into shapes, in part related to water content, and its ability to withstand firing. Firing produces two essential transformations in the raw material: It renders the vessel rigid and eliminates porosity so that water, food or other contents will not leak through. These changes are due to elimination of water content and to fusion of minerals within the clay, especially silica. The texture and color of the final product is determined by the chemistry of the clay and, of greatest importance, the temperature of firing, and the chemistry of the atmosphere in the kiln. Even in the Neolithic, potters had learned some simple ways to manipulate temperature and the atmosphere within the kiln by controlling the amount of air circulation during firing. When the supply of oxygen is unrestricted, an oxidizing atmosphere is created, giving the end product a red or yellow appearance. Conversely, with restricted air entry, a reducing atmosphere is created and the material tends to be grey. Allowing the air to become smoky creates a black color (SCC VI:6 63). The latter was probably easy to discover but other effects of airflow on color would not have been self-evident and must have resulted from observation of accidental effects that were then produced deliberately. Over time, the ability to control the appearance and characteristics of the final product were greatly refined. Kilns were specifically designed, for example, to produce down-drafts or cross-drafts. Control of temperature was critical and difficult because firings might last as long from four to ten days. Not only this but for optimal results, there had to be a sequence of temperatures.

Though it was common for firings to fail to produce proper pieces, when they did succeed, the technical quality of the wares was superb. That these complex processes worked at all indicates that very careful experimentation must have been done. This did not lead to a general conception of verification, nor did it lead to materialistic interpretation of the reasons for success or failure, such was inhibited by the standard cosmology. Social snobbery was also a factor inhibiting the development of theoretical understanding; as in many other cultures, the elite looked down upon those who worked with their hands.

The literate had always been concerned to maintain their power over those who actually had to do the physical labor of society. Second, Chinese literati were ambivalent regarding luxury goods and lavish lifestyles generally. Laozi, Confucius, Zhuangzi, and many later philosophers praised men, usually mythical, who wore rustic clothes and ate off simple wooden boards. As has often been pointed out, simple living was often an affectation. Thus the famous Suzhou "Garden of the Mender of Fishing Nets" is one of the most lavish in the world. It was built not by a humble worker but by a high government official. This tendency for the simple life to transform into a luxurious simulacre is hardly unique to China. One way the pretended disapproval of luxury goods was expressed in disdain for those who crafted them. Underlying the affectation of simplicity in living was a pervasive anxiety, never absent in agricultural societies, that if too few worked the land, there would not be enough food. It was assumed that production was fixed. The notion of progress was generally lacking in traditional societies and so improved productivity, though it did occur during the history of Chinese agriculture, was not relied upon. As already emphasized, the Confucian official class, schooled in the classics, saw return to earlier ways of life as the solution to modern problems. Poor harvests were as likely to be blamed on ritual impropriety as on farming methods. Technology did advance but in spite of the dominant ideology. This lack of belief in progress no doubt reduced one of the most important motives for scientific and technological research—the desire to solve problems.

China was ambivalent about the propriety of devoting resources to production of luxury goods for the rich, but of course production flourished because the rich were willing to pay for them. Not only from an economic perspective but from a spiritual one there is an inherent contradiction, one that recurs throughout human history and is irresolvable. The great spiritual teachers have always commended the simple life with few possessions and limited wants. Yet for many, one of the strongest evocations of spiritual feelings is contemplating objects of rare beauty. Modern societies partly resolve this by having the most valuable objects on display in museums. Lavish buildings such as Buckingham Palace or the Forbidden City (the former palace of the Qing emperors) are open to the public, at least some of the time. There was nothing like this in China. Collectors usually kept their art objects out of sight except for private enjoyment or to show to close friends. Temples however did have lavish objects on display, but other art was only for the privileged. Thus secular art, which included most porcelain, could not claim any public benefit.

Whether because of sumptuary concerns or not, in China technology does not seem to have been a major force in artistic innovation. While it clearly was for ceramics and for a few other arts such as cloisonné, such difficult crafts as jade carving continued to be executed laboriously with hand tools. When European clocks were brought to China by the Jesuits, they attracted considerable interest; the Forbidden City still houses the enormous number of elaborate clocks collected by the Qing emperors. China was not immune to the aesthetic appeal of mechanical technology but had not developed it indigenously. Instead its energies were devoted to creating some of the most exquisite handmade objects ever produced. To many Asian art has a spiritual quality of a sort not found in Western art. This is obviously a matter of individual taste but here it could be argued that China had just enough technology to create objects of the highest possible quality, yet not so much that technical aspects predominated.

## THE GODS OF THE KILNS

When one admires a beautiful collection of Chinese ceramics in a museum, it is sad to reflect that one motivation for producing wares of such perfection was the possibility of severe punishment for workers if firings failed. Supervision was by eunuchs and government officials who may have known little about the actual processes but would have been quick to punish nonetheless. Given that the results of firing were always uncertain, there must have been continual fear on the part of the potters and kiln workers. As with other Chinese occupations, the ceramic workers developed their own gods to help attain a good outcome from their efforts. Occupational gods are hardly less popular in China today with special gods for a wide variety of professions including opera singers, construction workers, taxi drivers (one of whom is Mao Zedong), prostitutes, gangsters, and police. The latter two occupations, intriguingly, invoke the same god, Guangong, once a real general in the Three Kingdoms period.

This practice of having gods dedicated to the worldly and spiritual needs of particular ways of life is a widespread religious phenomenon. Christian saints often protect particular activities or nationalities, for example. In India, there are occupational gods as well, also including ones for thieves and prostitutes. In Hinduism, deities are more often associated with philosophical abstractions than in China, until the arrival of Buddhism brought supernatural beings of Indian character into China. Arguably the most philosophical of the Chinese gods was Laozi, the legendary author of the *Dao De Jing*, who was deified centuries after the classic attributed to him was composed. As a god, Laozi at first representing wisdom but eventually became a guide on the way to immortality. Some of the spiritual traits that came to be associated with Laozi were influenced by Buddhist notions of supernatural beings. Laozi never represented abstract forces as did Brahma, Vishnu, and Shiva who embodied arising, enduring, and cessation respectively. This transformation is not unprecedented. Religions often enlarge their appeal by accepting popular elements that in modern perspective seem superstitious rather than spiritual.

A common cliché states that Chinese are practical—but of course all cultures must be practical or they would not survive. It is fair to say, however, that indigenous Chinese religion was rich in practices intended toward practical outcomes. For the most part Chinese spirituality looks toward being rewarded in this life rather than in a putative next one. While Buddhism offers the possibility of rebirth into a Pure Land or entry into nirvana, the form most shaped by the Chinese worldview, Chan (Zen) tends to emphasize enlightenment in this life. The goal of most Chinese devotional practices was thus protection against human and natural dangers and improved living conditions in this world, defined as being rich, living a long time, and having plentiful (male) offspring. Propitiation, usually with food offerings, was prominent, Chinese always being careful to stay in the good graces of the powerful denizens of other realms. The supernatural entities, with which people felt a personal relationship, were those whose concerns were close to their own. Thus each occupation had its own gods and there were household gods as well. To appreciate the occupational specificity of the native Chinese gods, we must remember that life then was quite constricted in comparison to life in the modern West. Most waking hours were spent at work, with some respite for meals, family affairs, and festivals. Thus if belief in gods is inherent in the human brain, as cognitive religious science holds, it is quite natural to imagine gods in relation to the activities that occupied most of one's life.

The gods of the potters are covered briefly by Kerr and Wood (SCC 2004:166f, 206, and 566f). Two sorts are mentioned, which can be termed human martyrs and supernatural guardians. (There may well be other categories; scholarly study of popular Chinese religion has only recently attained respectability.) Many Chinese gods, especially protective ones, represent promotion to divine status of real figures who died from treachery or from suicide. (The term "promotion" here is appropriate, emperors at times promoted or demoted gods.) An example of the first is Guangong, still the protector of police and gangsters, a heroic general who appears in the quasi-historical narrative *Romance of Three Kingdoms*. Extraordinarily strong and brave, a no-nonsense military man, he was captured by deceit and executed. Another is Zhong Kui, whose picture was posted on doors on New Year's to frighten away evil spirits. The legend is that Zhong Kui, having failed the civil service examinations, committed suicide by banging his head against a staircase in the imperial palace. Later, when the Tang emperor was frightened by ghosts in a dream, Zhong Kui suddenly appeared in the dream and scared them away. Both of these deified mortals died bravely, though not for any clear principle. In China and elsewhere many religious figures deemed to have great salvific power died under similar circumstances. Such deaths are related to the archetype of human sacrifice, a real practice in China until the Zhou. The same archetype is very prominent in Christianity, not only with the crucifixion of Jesus, but in the lives of the early saints, nearly all of whom died under torture. The ideas that the dead still exist somewhere and have power to help the living was near-universal until the advent of the modern worldview.

There were at least two martyr gods worshipped by potters; both ended their lives by leaping into the kiln fire. Death by fire was not only an obvious mode of suicide for a kiln worker, but also religiously significant. In our time, when fire is mostly a domesticated luxury, we forget how dramatic is its raging energy. In Hinduism the god of the fire, Agni, was initially the most important of the Vedic deities; tending the sacred flame was the highest duty of Brahman priests (Danielou 1964:63–66). The most profound transformations apparent to the ancients were those produced by fire. Not only does cooking greatly alter food, fire completely deprives many material objects of their structure and leaves undifferentiated material. In pottery manufacture, the transformation of mushy clay into objects that are hard, impermeable, and aesthetically appealing must have seemed in part supernatural. The risk of falling into the kiln must have been on workers' minds constantly and so it is not surprising that their gods were martyred in this way.

Kerr and Wood (SCC 2004:206) note that an early version of the story of a potter sacrificing himself by jumping into the fire was a fourth-century text in which the martyr was the pottery official under the legendary Yellow Emperor. In another version, it was the Yellow Emperor himself who leapt into the fire. The origin of such myths is notoriously difficult to determine and such is their resonance that many variants develop as the tales circulate. In a later Ming version, a potter named Dong leapt into the fire to aid his fellows, who were being cruelly punished by their eunuch supervisors for their failure to properly fire difficult vessels. After Dong's sacrifice, the vessels fired perfectly. Here is a clear motif of human sacrifice, though voluntary. At least some Ming emperors personally made offerings to ceramic gods, indicating how important elegant ceramics were to the imperial courts.

This tradition continues in another myth, this one related to the difficulty of making copper-red glazes, which required extremely precise conditions for successful results. During the Ming, the kilns of Jing De Zhen could no longer produce this glaze. A potter committed suicide by leaping into the kiln. Later another potter dreamed of his deceased colleague who related the secret of making the glaze. The similarity to the story of Zhong Kui is apparent but here the benefit is practical rather than supernatural. It is tempting to speculate that the surviving potter somehow intuited or recalled the method during a dream, a not unusual event in creative work. However, the elements of the story are conventional and it is equally plausible that the story was invented to enhance the importance of the rediscovery. Here too, there is a shamanistic element in that solving a problem in this world requires a visit to another realm, in this case via a dream.

Another technological improvement was said to have been transmitted via a different kind of dream. Another legendary potter, Lin Bing, had a kiln that was too small for adequate production. During a dream, a goddess appeared to him and recommended that he construct a kiln shaped like her female body, with a rounded dome representing her abdomen and two smaller domes analogous to her breasts. This design was widely adapted in the De Hua region. Attractive goddesses are not unusual in Chinese mythology, and there is no reason to imagine that alluring female figures were any less common in men's dreams than they are now. This dream is somewhat racier than those in usual Chinese accounts. This may be due to the kiln workers being more open about their sexual thoughts than the often sanctimonious Confucian literati. This goddess belongs to the category of supernatural guardian since her myth does not include anything regarding an earthly life.

How the kiln actually came to have this nonintuitive but efficient shape is uncertain. Since the design must have been arrived at by formulating a hypothesis and testing it by constructing the first kiln of this design, it seems unlikely that the designer was really inspired by his appreciation of the female figure. More likely the explanation came after the design was complete. Significantly, these mythical explanations of origins were of more interest than the actual creative process, which was usually forgotten.

Inspiring or inspired figures in myth are often female, as are the symbolic muses in Greek and later Western literature. This link between femininity and creativity is obviously related to the obvious fact that it is women who give birth but probably also to male psychology for which women are mentally stimulating, sexually and in other ways as well.

In the ancient world invention was usually explained by invention of myth in contrast to our own era, which is fascinated by artistic and scientific creativity as mental processes. For us, explanations are based on what we think is the nature of the mind. In traditional societies, this notion of mind did not exist and so supernatural explanations were employed. Now these are often psychologized as in Jung's famous dictum regarding the reduction, begun by Freud, "In our time the gods have become [psychological] diseases."

It is tempting to conclude that supernatural explanations for discoveries, by obscuring the human origin of invention, tended to inhibit deliberate efforts to make them. Yet in the technology we are considering here China excelled and so perhaps myth supported creativity in its own ways. Nonetheless, the preponderant effect of mythological explanations, such as those in the *Da Zhuan* of the *Yi Jing*, must have been to inhibit actual research.

With potters' deities as with those of smallpox we find empirical validation of chemical techniques coexisting with elaborate religious beliefs and practices. That experiment could reveal the best ways to form each type of ceramic, and even create new ones, was not taken to mean that supernatural help was unnecessary. Though, as we have seen, China did have skeptics, there was never an explicit conflict between empiricism and religion. While modern science tends to be blamed for a materialist ideology that regards human existence as a mere random event, empiricism does not depend on any particular belief system. However, religious belief tends to set limits for empiricism for two reasons. First, specific doctrines such as yin-yang or the three gunas are not falsifiable, nor was there any need for proof beyond the writings of the ancient sages. Second, when religious doctrines, such as creation accounts, are in principle refutable, there tends to be avoidance of any train of thought that might challenge them. The latter is more characteristic of the religions of the Book, where the very existence of written scriptural authority creates anxiety that part of it might be questioned. While the major Asian religions all have canonical texts, there is less commitment to each word of these as the word of God and thus infallible. With respect to Chinese and Indian cosmology, there never seems to have been much concern with their literal truth. They had personal and social explanatory power and that was enough.

I was recently told by an American Daoist teacher that Chinese scientists had developed a device to detect qi. The notion of measuring qi with an instrument is clearly modern; it is an apologetic to justify qi as scientific. In traditional China, though compass needle movements were attributed to qi, a device to verify qi would never have been thought necessary. Qi was there, whether measured or not. It is in no small part because they have been less concerned with the material reality of their cosmology that Asian religions have not been greatly threatened by the entry of Western science. Thus empiricism flourished in China side by side with its metaphysical cosmology. There are some recent exceptions, such as attempts to demonstrate that the Hindu creation of myths are supported by science, but these appeal only to a small fringe (Cremo 1998). What is most striking about this and similar efforts is their failure to attract much interest among Hindus.

## CAN TECHNOLOGY BE SPIRITUAL?

In the modern world, technology is blamed for many of the perceived ills of society including petroleum and global warming, atomic weapons, toxic waste, trans fats, drug side effects, and many more. Our dependence on technology is often proclaimed to be a weakness, as if we could all grow our own food. At the same time it is technology that most determines the nature of our lives. In New York City, where I live, life would not be possible without electricity, telephones, subways, elevators, and food preservation methods, to name just a few. Part of the criticism of technology stems from the persistence of the notion, apparent in the earliest Chinese literature, that humanity was somehow better off in archaic times. A variety of movements have taught this idea, from the Puritans who first settled in New England, to new age advocates of the simple life, to charismatic founders of communes, to the genocidal Khmer Rouge who killed about one fourth of the population of Cambodia, including anyone who wore glasses or could read. In China, many of the canonical texts, including the *Dao De Jing*, the *Lun Yu*, and others, advocate returning to previous forms of society as the cure for social problems.

As a physician, I spend much of my time applying technology to relieve human suffering. Whatever the faults of modern technological medicine, its ability to return people to active participation in life is of unquestionable spiritual value. Other nonmedical examples abound. Cell phones can be annoying but can be lifesaving for those who live alone. The glib assumption that technology is somehow unspiritual must be set aside. Much technology has aesthetic value beyond its practical use; ceramics are a case in point. For many, aesthetic pleasure is spiritual in that it gives a sense of meaning to one's life. In China, ceramics, though used to a limited degree in temples and religious observances, mainly served as prestige goods. Imperial patronage was a major impetus to advancing its technology.

However selfish the motives for production of these expensive luxuries, the kilns provided employment and helped enhance China's image in the world.

While we tend to think of conspicuous consumption as the result of capitalism, archeological research has clearly shown that use of objects to establish status is at least as early as the upper Paleolithic (between ten thousand and thirty-five thousand years ago) (Hayden 2003:90, 130). The most significant equivalent of modern conspicuous consumption was lavish burial. So that the rich would have what they needed in the next world, expensive objects, such as manuscripts, jewelry, and bronze vessels were buried with them. In a time when possession of a tool or a cooking utensil marked one as wealthy (still the case in much of the world), burial of such goods was exceedingly wasteful. Worse, during the Shang, the king's officials and servants were buried alive with him. This was not only cruel but also extremely wasteful of human resources in an era when population sizes were marginal and those with experience of government were few. While it often helped stabilize the successor's hold on power, it deprived him of skilled officials. We of course benefit from these practices, which have preserved many objects of aesthetic and archeological interest.

The nominal reason for extravagant burial practices was to provide for the deceased in his next life. (Women were provided fewer burial objects, though they gradually increased over time.) Anthropologists look beneath the surface for explanations of social behavior, especially when it is irrational, and emphasize that common to all such rituals is the desire to display wealth to enhance status and demonstrate power. Sumptuary burials were nearly universal in all but the very earliest prehistoric societies. The desire for prestige articles has been and still is a major impetus for the advancement of technology. We like to think that spirituality, at least in the best sense, is indifferent to status and prestige. However there is often a close association between the cost/prestige value of an item and its spiritual value. If we think of medieval cathedrals, their splendor may express devotion to God but it also shows off the wealth of the society and the religious intuition. Similarly for colossal Buddha statues, the cave murals of Dunhuang, the temples of Ajanta and Ellora, the Taj Mahal, and many other monumental religious sites.

In China there were sumptuary rules regulating display but because of the inherent human love of showing off, they were not effective. Such rules arise because for some, such as Laozi and Confucius, spirituality means living with no more than one actually needs. In general in China display was more a matter of quality than quantity, the emperor's court and some homes of the very wealthy excepted. Thus fine porcelain was highly valued despite these objects being small and delicate; such objects demonstrated aesthetic refinement at least as much as wealth. Similarly, scrolls of calligraphy and painting were usually rolled up and brought out only as a show of respect for close friends or important guests.

If technology creates lavish goods affordable only to a few, it also makes plainer ones available to the masses. Many of these were spiritual in nature from terra-cotta amulets to printed sutras. It is not by accident that the first printed book in Europe was a Bible, not the earliest surviving printed work in China, the Buddhist Diamond Sutra. Technology makes spiritual experience both more widely available, as with printing, and more intense, as with more skillfully cast metal images. Application of technology to religion continues today. In Buddhist events in Hong Kong, audiences are sometimes given iPod-size electronic devices that endlessly play recorded mantras, usually *Amito Fo*," "Homage to Amida Buddha," over and over. On a grander scale, several Asian countries continue to compete to build the tallest Buddha statue. The first such effort was at Fuo Guan Shan outside Gaoxiong in Taiwan. Not to be left behind, the Bao Lian Si (Precious Lotus Temple) in Hong Kong raised the necessary funds and constructed an even taller one on Lantau Island next to Hong Kong. Not to be outdone, China has begun to restore a 1280-year-old 71-meter-tall statue in Leshan, Szechwan, at a cost of 30 million dollars. This last project has more spiritual appeal since it restores an ancient work. In this case of the growing Buddhas we can see the same point that was made with respect to porcelain: technology, status display, and spirituality are interwoven and any separation is artificial. Even those most disdainful of material wealth must recognize that spirituality would be impossible without such objects as scriptures, sacred objects, and religious images.

# Chapter 10

＿＿

# Asian Spirituality and Science in the Modern World

This book has concerned itself with science and spirituality as it developed in Asia before the advent of significant Western influence. We have seen that science and religion as two modes of thought were closely associated and that there was no overt clash between science and religion. Some expressed skepticism regarding the validity of divination or the existence of supernatural beings but not to the degree of completely rejecting the fundamental metaphysical paradigm of their culture. Indeed, the cosmologies of China and India were such powerful explanatory systems that, despite the occasional skepticism just referred to, they remained largely intact until the incursion of Western culture. During the late Qing dynasty, there was a rise in empiricism, though more pronounced in textual studies than in science (Elman 2001); however alternatives to the traditional cosmology were never devised. It was not so much questioning from within that led to abandonment of the theory of yin-yang, the five phases, qi, and li, but the stark power, both military and civil, that empirical science bestowed upon the Western powers. The spiritual strength of the Indian and Chinese religions was largely mental and could not match the material effects of science. The rigorous self-discipline of martial arts such as *qi gong* lost its credibility when found to be completely ineffective against an enemy armed with guns and explosives. Since training in qi gong was based on the correlative metaphysics of yin-yang and qi, its failure resulted in disillusionment regarding the ancient beliefs. There was of course some resistance to Western science. One form it took earlier in the Qing was the claim that this seemingly new knowledge from outside China had actually first been discovered by the ancient sages and thence passed to the West (Elman 2001:116). This preserved the idea that truth was to be found in returning to the wisdom of the ancients, but became less and less plausible over time as Western science advanced.

Momentum for change had been building up even before Puyi, the last emperor of the Qing dynasty, was easily toppled in 1911. Many of the educated young, gathered in the May Fourth movement of 1919, advocated adaptation of modern Western ways, particularly science and democracy, as the only way forward for China. The civil service examination system based on the Four Books had been abandoned in 1905, allowing the educational system to be broadened. Modern scientific institutes were founded early in the twentieth century. Peking Union Medical College (PUMC) was established in 1906 with both governmental and religious organizational support and, a few years later, with significant aid from the Rockefeller Foundation. After China's opening to the West in the 1970s, the government placed priority on developing PUMC as the leading medical institute in China. In a bow to tradition, the old romanization of "Peking" rather than "Beijing" was retained. While politics and connections still play a role in faculty selection, the government's commitment to scientific medicine is apparent. Institutes for other sciences have been founded and many Chinese study science, medicine, and other fields in European and American Universities.

Ironically, one of the Western ideas that had great impact on life in Asia was Marxism, paradoxically turned against the West where it had originated. On the other hand, the other seminal thinker of the twentieth century, Sigmund Freud had virtually no influence. Nor has more recent Western psychiatry made much inroad. Asian adaptation of Western ideas has been selective. All the Asian spiritual traditions offered highly developed methods of self-examination but the notion of blaming one's parental upbringing for one's troubles is incompatible with the core value of filial piety.

Neither Western colonization, nor the complete acceptance of modern science has driven out traditional ways of thought in India or China. While education no longer is based on the *Vedas* or the Confucian classics, devotional rituals and divination continue to be widely practiced. Selection of auspicious days for important events continues to be nearly universal. These beliefs and practices may be kept in a separate mental compartment from modern technical knowledge such as economics and science, but they have not been abandoned. Yet the traditional cosmology no longer forms the basis of intellectual discourse. Few, except scholars, have detailed knowledge of the traditional cosmology such as the three gunas of five phases. On the other hand, the less abstract metaphysical principles such as prana and qi still are used to explain yoga, meditation, and martial arts—practices which are more popular than ever. The doshas of Ayurveda, like the various forms of qi of traditional Chinese medicine

(TCM) are still the theoretical basis of traditional medicine. The ancient texts such as the *Vedas* or the Four Books are not often read but others such as the *Bhagavad Gita, Dao De Jing,* and *Yi Jing* have a wide readership in modern translations. In India and the West beginning in the late nineteenth century, philosophers attained great prominence by teaching the traditional spirituality in ways accessible to modern urbanites and suburbanites. These include Ramakrishnan, Vivekananda, and Sri Aurobindo. Although they have been criticized for incorporating some Western elements, their following demonstrates that the ancient traditions continue to inspire. Yoga, though usually stripped of much of its spiritual content, is practiced everywhere. In China the situation has been less happy. The Communist "liberation" under Mao Zedong in 1949 led to almost total suppression of independent thought and, during the Cultural Revolution, efforts to obliterate the religious traditions. No Chinese spiritual figure has achieved worldwide admiration as have Mohandas Gandhi or the Dalai Lama. While Confucianism, with its hierarchical view of society and sexism, has not been particularly attractive in the modern world, Daoism and Buddhism are flourishing anew in the liberalized post-Mao era.

The persistence of the older beliefs and practices has not inhibited development of science in China or India. Both have their share of Nobel laureates and those born in both countries are prominent in European and American science. It is my impression that scientists and physicians from these cultures are less likely to interest themselves in traditional metaphysics or such practices as feng shui, but this is far from absolute. In both countries, the old practices for ensuring success continue to be popular. A friend of mine who is a prominent gynecologist in India has her apartment filled with statues of Ganesh, the Indian god of success. I have a small antique image of him in my office and Indian patients sometimes comment on my prudence in invoking Ganesh in this way. I don't think these practices are always taken entirely seriously but they are not derided either. While hard core skeptics are annoyed by the persistence of anything remotely supernatural, the world would be a much poorer place if all that cannot be empirically verified were excluded from our lives. At the same time, it would be foolish to rely on supernatural aid for practical matters such as building safe bridges or curing infections.

Many of those in the modern West who feel a lack of spiritual roots find the continuity of tradition in Asia attractive. We cannot go back to the traditional world but we can learn from it and extract spiritual ideas and practices that enrich our lives. It has been relatively easy to separate core ideas from Indian and Chinese philosophy from the supernatural beliefs with which they coexisted comfortably in their original form. This demythologization has occurred not only in the West but also in India and China as well. That spirituality in Asia did not have a history of overt conflict with science and modernity generally has made it accessible to many who find aspects of Christian history troubling. To be sure, one reason for this is that Westerners are far more aware of the failings of our own religious institutions than those of Asian civilizations that were almost inaccessible until recently.

Asian religion has not always been presented to the West in demythologized form. One of its great popularizers was Helena P. Blavatsky (1831–1891), founder of the Theosophical Society, who had many followers. Her theosophy was an eclectic amalgam of Hinduism, Buddhism, and Western esotericism with great emphasis on reincarnation, channeled masters, séances—clearly faked—and the like. The Theosophical movement borrowed from Asian religion in superficial ways to give its doctrines and teachings the allure of exotic antiquity. A century later, in the sixties, this fascination with the East became even more widespread. From the point of view of scholars, this popularization has been a mixed blessing. It has made not only Asian spirituality available to millions but also simplified, distorted, and conflated its rich variety. The Asian religion of a contemporary weekend seminar is not that of the pandits or sages. Even in Asia, the traditional spirituality is usually presented in simplistic versions. This however is not new. In the past, it was only a few who were familiar with the difficult texts now studied by scholars.

Many hope that Asian spirituality can offer methods for alleviating what is perceived as the widespread spiritual malaise of modern life. Science is often blamed for contributing to this by depicting the universe as dead objects floating in vast, uninhabitable spaces. It is often suggested that this universe in which humanity has no special place is a cause of contemporary angst. On the microcosmic level, medicine is blamed for treating patients as collections of parts rather than as whole individuals. Also, technology derived from science has had adverse environmental effects that have been difficult to bring under control. The history of Asian science and spirituality shows that similar problems existed then, but the response was more often ritual activity than practical efforts. There tended to be at least as much fear of supernatural calamities as of natural ones. In earlier times as now, the factors limiting conservation were basically social and political.

Many sense something spiritual in the universe and feel that this sense is expressed well in Asian spiritual traditions. The experience of an underlying harmony in the universe analogous to what the Chinese referred to as the Dao and Hindus as the Atman does not in any way contradict science. Though the cosmologies describe fantastic realms inhabited by supernatural beings, these aspects of Asian religion have seemed optional to many moderns. Supernatural beings can be conceived as symbols or archetypes rather than as having an existence outside the mind. Similarly, doctrines such as rebirth can be interpreted metaphorically as expressing the interrelation of all sentient beings. Some accept these tenets literally but most see them as optional.

Buddhism has been the Asian spiritual system that has attained the greatest following in the West, due to its emphasis on the relief of suffering and on the actual practice of meditation. While spiritual exercises comparable to meditation were practiced at times in the West, they were never widespread. A great appeal of Asian spirituality is the strong tradition of meditation involving both body and mind, which many prefer to the more passive experience of sitting in church listening to sermons or hymns. Zen, presented by D. T. Suzuki, Alan Watts, and others as tending toward atheism and iconoclasm, was the form of Buddhism that first attracted interest, although its association with the counterculture had little relation to its social role in Japan. In the past two decades Vajrayana (Tibetan) Buddhism seems to have attained wider appeal. This change from Zen, a form of Buddhism which gives relatively little place to the supernatural, to one with a plethora of deities and extensive rituals suggests that Buddhism in the West is following a trajectory reminiscent of religious development in both China and India where the relative austerity of early Buddhism and Daoism evolved into something quite different. In China, the philosopher Laozi became a god and devotional practices tended to put the philosophy in the background. Meditation involved very elaborate circulation of qi throughout the body. In Buddhism, the Mahayana offered thick tomes of abstruse philosophy and a plethora of deities. In Vajrayana, real meditation could not be until one had completed *ngondro*, a set of difficult preliminary exercises including performance of 108,000 prostrations. The lack of conflict between Asian religion and science seems to be the case in the West also. A high proportion of Buddhist practitioners have advanced academic degrees and few seem to find much difficulty in combining Buddhism with modernity.

It is curious that the initial appeal of Zen as direct practice without need for study or ritual participation evolved into fascination with the more baroque Vajrayana. One reason is historical: only after the Chinese annexation of Tibet did its teachings become available outside their homeland. The leadership of the present Dalai Lama has played an important role in attracting Westerners to Vajrayana. The appeal of this form of Buddhism also suggests that for many, even scientists, the purely intellectual aspect of religion is not completely satisfying. Vajrayana holds that at the highest level of understanding, the deities are part of one's own mind. Nonetheless they are usually spoken of as if they are real. This amounts to having one's spiritual cake and eating it too. There is opportunity for devotion toward gods without having to actually believe in them, one that fits well with Jung's view of gods as archetypes inherent in the human mind. Western religions do not offer anything like this. It seems that many who cannot believe in the supernatural still want it to be there. Thus in some ways we are akin to the Chinese kiln workers who applied empirical methodology with great skill while invoking the aid of gods dedicated to their craft.

Scholars tend to be irritated by the covert modernization in many popular writings on Asian philosophy, religion, and healing. Yet, to make practical use of the spirituality of the past we must adapt it to our own needs. I suggest that both approaches are compatible. There is great value in attempting to understand how our fellow humans thought in the past; this is not contradicted by modifying these ways of thought to fit our very different worldview. We can talk of yin and yang while recognizing that we are not using the terms exactly as the Chinese did two thousand years ago. We can think of karma as a moral principle inherent in the universe without believing in the heavens and hells of the Mahayana. Modernization of Asian spiritual practices and beliefs has been relatively straightforward, although many do not recognize how the traditional forms have been altered. For example, the teaching, widespread in traditional Buddhism, that it is not possible to be enlightened in a female body, is quietly ignored.

## ASIAN SCIENCE IN THE MODERN WORLD

While Asian religion has found a place in the modern world with surprising ease, traditional science, particularly medicine, raises more complex issues that remain to be resolved. Spirituality is ultimately personal—if one finds a set of beliefs satisfying then they are correct for oneself. Religion exists as social structures but in the modern world these are generally voluntary, though they can exert pressure on members. Religious ideology stops being a purely personal matter when it induces guilt or fear to manipulate followers or when used to incite harmful behavior, such as terrorism. The spiritual traditions we have been discussing have rarely favored violence, though there have been regrettable lapses, such as the Japanese Zen establishment allowing itself to be used to help justify the imperialism that led to the Pacific War (Victoria 1997). With a few such exceptions, Asian spiritual traditions have emphasized nonharming; this is part of their attraction for moderns.

Most traditional sciences, such as cosmology and astronomy, are now of historical interest. The degree of accuracy of ancient astronomical observations, to give but one example, is without practical consequence in the modern world. The medicine of China and India however are experiencing a revival, both in the West and in their own countries. Their safety and efficacy are thus of social concern. Since these medical systems are based on metaphysics and anecdote, their texts provide no evidence to determine safety and efficacy that meets current standards of evidence-based medicine. There is some evidence for acupuncture in chronic pain but almost none for herbs in any condition. Even those with known pharmacological activity, like the stimulant *ma huang* (ephedra), do not have any clear therapeutic value. (Sale of this has been banned in the United States because it can be used to make illicit drugs.) Safety is entirely uncertain; not only because of lack of systematic knowledge about the herbs themselves but also because many of those sold are adulterated with poisonous substances. The status of traditional Asian medicine in the West is thus ambiguous. The ethic of tolerance for other cultures may conflict with the expectation of Westerners that the government ensure that consumer products are completely safe. This conundrum awaits resolution.

Since the publication of Fritjof Capra's *The Tao of Physics* and Gary Zukav's *Dancing Wu Li Masters*, a spate of popular books have asked us to imagine that the ancient Chinese and Hindus somehow arrived at the discoveries of modern quantum physics or the structure of DNA (Capra 1991; Zukav 1979). Even those who should know better have succumbed to this temptation, as even Joseph Needham did when suggesting that yin and yang are similar to positive and negative charges in electricity. It only needs to be pointed out that no amount of theorizing about yin and yang made possible the development of the light bulb. These claims are anachronistic in the extreme and serve only to impede understanding of both physics and Asian religion.

We have seen that in China and India, correlative cosmology probably inhibited the progress of science, not because of any overt conflict but because metaphysical explanations satisfied curiosity sufficiently to limit speculation outside the received system. Some were aware of the limitations of the standard cosmology but in the absence of a philosophy of empiricism, were unable to find a way to go beyond it. Yet empirical discoveries of great importance were made and refined; the examples given have been smallpox vaccination and ceramic technology. In both cases, empiricism and religion functioned side-by-side. In contrast, much contemporary work on science and religion seeks to merge the two, as if somehow science will confirm religious beliefs. This seems extremely unlikely to happen now and it did not happen in traditional Asia. However there is no indication that either science or religion will eventually drive out the other. They will continue to live together in an unstable equilibrium for the foreseeable future.

# Primary Sources

## —1—
## Pankenier David W. In Mair, Victor H., Steinhardt, Nancy S., and Goldin, Paul R., eds. 2005. *Hawai'i Reader in Traditional Chinese Culture*. Honolulu: University of Hawai'i Press, 24–26.

The first excerpt is from the *Zhouli*, one of the five Confucian classics. The remainder is from China's most famous historian, Sima Qian, and describes the calendar reform undertaken during the reign of Emperor Wu (141–87 B.C.E.) of the former Han that took 104 as the base epoch. Though the new calendar was based on observation that had revealed the inaccuracies of the older one, the passage demonstrates that these observations were immediately fitted into the correlative system, for example, with musical notes. There is also the implication that need to correct the calendar has occurred only because the writings of the (mythical) Yellow Emperor were lost. This amounts to an apology for basing the new system on observation rather than on ancient texts. That natural phenomena can be independent of human action is not recognized. Sima Qian's treatment is characteristic of the literati trained in the classics but not in science.

[ ... ]

Ever since the people have existed, when have successive rulers not systematically followed the movements of sun, moon, stars, and asterisms? Coming to the Five Houses (Huang Di, Gao Yang, Gao Xin, Tang Yu, Yao-Shun) and the Three Dynasties, they continued by making this [knowledge] clear, they distinguished wearers of cap and sash from the barbarian peoples as inner is to outer, and they divided the Middle Kingdom into twelve regions. Looking up they observed the figures in the heavens, looking down they modeled themselves on the categories of earth. Therefore, in Heaven there are Sun and Moon; on Earth there are yin and yang; in Heaven there are the Five Planets; on Earth there are the Five Phases; in Heaven are arrayed the lunar lodges, and on Earth there are the terrestrial regions.

[ . . . ]

When it came to the accession of the Sovereign, he summoned the specialists in recondite arts, Tang Du, to reapportion the heavenly sectors (lunar lodges), and Luoxia Hong of Ba to convert the angular measurements into a calendrical system. After that the graduations of the chronograms conformed to the regulations of Xia. Thereupon, a new regnal era was begun, the titles of offices were changed, and [the emperor] performed the *Feng* sacrifice on Mount Tai (in the summer of 110 B.C.E.). Accordingly, [Emperor Wu] issued an edict to the Imperial Scribes, which said, "Recently, We were informed by the responsible officials that the astronomical system still has not been fixed. Having widely solicited advice about bringing order to the stellar graduations, they have been unable to resolve the matter.

"Now; We have heard that in antiquity the Yellow Emperor achieved perfection and did not die. He is renowned for having examined into the graduation of the stellar regions] and verification [of their portents], determined the clear and turbid [notes of the musical scale], initiated the [sequence of the] Five Phases, and established the division and number of the [24 fortnightly] *qi*-nodes and their [associated] phenomena. Thus, [concern with such matters] goes back to high antiquity. We deeply regret that the ancient writings are lost and the music abandoned, so that We are unable to perpetuate the brilliance [of antiquity], [now that] the accumulation of days and temporal divisions properly conform to the subduing of the influence of Water [of the Qin]."

# —2—
# Lynn, Richard John, trans. 1994. *The Classic of Changes: A New Translation of the I Ching as Interpreted by Wang Bi.* New York: Columbia University Press, 142–148; 47–56.

The *Yi Jing* (*I Ching* in the older Wade-Giles transliteration) was one of the most important canonical texts of traditional China. The earliest portions, sometimes referred to as the *Zhou Yi*, are thought to have been composed in the Early Eastern Zhou dynasty, about 900–800 B.C.E. This consists of the famous hexagrams, titles for each, a cryptic explanation, and a brief text for each of the six lines of the hexagram. These early texts are laconic to an extreme and so most Chinese editions, as well as those in Western languages, insert commentaries into each Zhou text taken from the so-called Ten Wings. These were appended in the late Warring States or early Han.

Two selections are included. The hexagram selected, "*kun*," representing earth, receptivity, and yin, which gives a sense of the obscurity of the early text and the ingenuity with which meaning was inserted into it by later commentators. The second is an extended passage from the most important of the "Ten Wings," the *Da Zhuan* or "Great Commentary." The *Da Zhuan* is the foundational text of Chinese correlative metaphysics; almost all important themes that dominated Chinese thought for the ensuing two thousand years can be found in the selection here, including the correlation of natural events with those in human society and the equation of natural laws with ethical ones.

This translation is the recent one of Richard John Lynn. While less well-known and less literary than the Wilhelm-Baynes version, it is more accurate and includes commentary by Wang Bi (226–249), the boy-genius of Chinese philosophy whose commentary largely determined interpretation of the *Yi Jing* for the following two millennia. In Lynn's text, reproduced here, commentary on the second hexagram, "Kun" by Wang Bi is enclosed in curved brackets {}. In the section from the *Da Zhuan*, curved brackets are used to designate commentary by Han Kangbo (d. ca 385), whose approach to the classic was essentially derived from Wang Bi's.

*Kun* [Pure Yin]
(*Kun* Above *Kun* Below)
[ . . . ]

## COMMENTARY ON THE IMAGES

Here is the basic disposition of Earth: this constitutes the image of *Kun*. {In physical form, Earth is not compliant; it is its basic disposition that is compliant.} In the same manner, the noble man with his generous virtue carries everything.

[ . . . . ]

## COMMENTARY ON THE APPENDED PHRASES

As Heaven is high and noble and Earth is low and humble, so it is that *Qian* [Pure Yang, Hexagram I] and *Kun* [Pure Yin] are defined.

The Dao of *Kun* forms the female.... *Kun* acts to bring things to completion.

*Kun* through simplicity provides capability.... As [it] is simple, it is easy to follow.... If one is easy to follow, he will have meritorious accomplishments.

When [the Dao] duplicates patterns, we call it *Kun*. As for *Kun*, in its quiescent state it is condensed, and in its active state it is diffuse. This is how it achieves its capacious productivity.

This is why closing the gate is called *Kun*.

*Qian* and *Kun*, do they not constitute the arcane source for change! When *Qian* and *Kun* form ranks, change stands in their midst, but if *Qian* and *Kun* were to disintegrate, there would be no way that change could manifest itself. And if change could not manifest itself, this would mean that *Qian* and *Kun* might almost be on the point of extinction!

All the activity that take place in the world, thanks to constancy, is the expression of the One.... *Kun* being yielding shows us how simple it is.

[ ... ]

## PROVIDING THE SEQUENCE OF THE HEXAGRAMS

Only after there were Heaven [*Qian*, Pure Yang, Hexagram I] and Earth [*Kun*, Pure Yin], were the myriad things produced from them. What fills Heaven and Earth is nothing other than myriad things.

[ ... ]

## COMMENTARY ON THE IMAGES

The frost one treads on becomes solid ice: This yin thing begins to congeal. Obediently fulfilling its Dao, it ultimately becomes solid ice.

[ ... ]

## COMMENTARY ON THE WORDS OF THE TEXT

When Heaven and Earth engage in change and transformation, the whole plant kingdom flourishes, but when Heaven and Earth are confined, the worthy person keeps hidden. When the *Changes* say "tie up the bag, so there will be no blame, no praise," is it not talking about caution?

[ ... ]

# Commentary on the Appended Phrases [Xici zhuan], *Part One*

As Heaven is high and noble and Earth is low and humble, so it is that *Qian* [Pure Yang, Hexagram I] and *Kun* [Pure Yin, Hexagram 2] are defined. {It is because *Qian* and *Kun* provide the gateway to the *Changes* that the text first makes clear that Heaven is high and noble and Earth is low and humble, thereby determining what the basic substances of *Qian* and *Kun* are.} The high and the low being thereby set out, the exalted and the mean have their places accordingly. {Once the innate duty of Heaven to be high and noble and that of Earth to be low and humble are set down, one can extend these basic distinctions to the myriad things, so that the positions of all exalted things and all mean things become evident.} There are norms for action and repose, which are determined by whether hardness or softness is involved. {Hardness means action, and softness means repose. If action and repose achieve normal embodiment, the hardness and softness involved will be clearly differentiated.} Those with regular tendencies gather according to kind, and things divide up according to group; so it is that good fortune and misfortune occur. {Thus similarities and differences exist, and gatherings and divisions occur. If one conforms to things with which he belongs, it will mean good fortune, but if one goes against things with which he belongs, misfortune will result.} In Heaven this [process] creates images, and on Earth it creates physical forms; this is how change and transformation manifest themselves. {"Images" here are equivalent to the sun, the moon, and the stars, and "physical forms" here are equivalent to the mountains, the lakes, and the shrubs and trees. The images so suspended revolve on, thus forming the darkness and the light; Mountain and Lake reciprocally circulate material force [*qi*], thus letting clouds scud and rain fall. This is how "change and transformation manifest themselves."} In consequence of all this, as hard and soft stroke each other. {That is, they urge each other on, meaning the way yin and yang stimulate each other.} the eight trigrams activate each other. {That is, they impel each other on, referring to the activation that allows change to fulfill its cyclical nature.}

It [the Dao] arouses things with claps of thunder, moistens them with wind and rain. Sun and moon go through their cycles, so now it is cold, now hot. The Dao of *Qian* forms the male; the Dao of *Kun* forms the female. *Qian* has mastery over the great beginning of things, and *Kun* acts to bring things to completion. {The Dao of Heaven and Earth starts things perfectly without deliberate purpose and brings them to perfect completion with no labor involved. This is why it is characterized in terms of ease and simplicity.}

[ ... ]

Therefore what allows the noble man to find himself anywhere and yet remain secure are the sequences presented by the *Changes*. {*Sequences [xu]* mean the succession of images *[xiong]* in the *Changes*.} What he ponders with delight are the phrases appended to the lines. Therefore, once the noble man finds himself in a situation, he observes its image and ponders the phrases involved, and, once he takes action, he observes the change [of the lines] and ponders the prognostications involved. This is why, since Heaven helps him, "it is auspicious" and "nothing will fail to be advantageous."

[ ... ]

The *Changes* is a paradigm of Heaven and Earth, {[The sages] made the *Changes* in order to provide a paradigm of Heaven and Earth.}, and so it shows how one can fill in and pull together the Dao of Heaven and Earth. Looking up, we use it [the *Changes*] to observe the configurations of Heaven, and, looking down, we use it to examine the patterns of Earth. Thus we understand the reasons underlying what is hidden and what is clear. We trace things back to their origins then turn back to their ends. Thus we understand the axiom of life and death. {The hidden and the clear involve images that have form and that do not have form. Life and death are a matter of fate's allotment for one's beginning and end.} With the consolidation of material force into *[jingqi]*, a person comes into being, but with the dissipation of one's spirit *[youhun]*, change comes about. {When material force consolidates into essence, it meshes together, and with this coalescence, a person is formed. When such coalescence reaches its end, disintegration occurs, and with the dissipation of one's spirit, change occurs. "With the dissipation of one's spirit" is another way of saying "when it disintegrates."} It is due to this that we understand the true state of gods and spirits. {If one thoroughly comprehends the principle underlying coalescence and dissipation, he will be able to understand the Dao of change and transformation, and nothing that is hidden will remain outside his grasp.}

As [a sage] resembles Heaven and Earth, he does not go against them. {It is because his virtue is united with Heaven and Earth that the text says: "resembles them."} As his knowledge is complete in respect to the myriad things and as his Dao brings help to all under Heaven, he commits no transgression. {It is because his knowledge comprehensively covers the myriad things that his Dao brings help to all under Heaven.} Such a one extends himself in all directions yet does not allow himself to be swept away. {Responding to change, he engages in exhaustive exploration but does not get swept away by illicit behavior.} As he rejoices in Heaven and understands its decrees, he will be free from anxiety.

[ ... ]

"Actually, how could there ever be an agency that causes the interaction between the polarity of yin and yang or the activity of the myriad things to happen as they do! Absolutely everything just undergoes transformation in the great void [daxu] and, all of a sudden, comes into existence spontaneously. It is not things themselves that bring about their existence; principle here operates because of the response of the mysterious [xuan]. There is no master that transforms them; fate here operates because of the workings of the dark [ming]. Thus, as we do not understand why all this is so, how much the less can we understand what the numinous is! It is for this reason that, in order to clarify the polarity of yin and yang, we take the great ultimate [taij], the initiator of it, and, in addressing change and transformation, we find that an equivalent for them is best found in the term *numinous*. Anyone who understands how Heaven acts will exhaust principle and embody change, sit in forgetfulness and cast aside the things in his care. As it takes the perfect void [zhixu] to respond perfectly to things, we equate this with the Dao. As it takes the complete lack of conscious thought to view things from the point of view of the mysterious, we call this the numinous. One who takes the Dao as resource and so achieves union with it derives his power to do so from the numinous but is himself more dark-like than is the numinous."}

[ ... ]

The sages had the means to perceive the mysteries of the world and, drawing comparisons to them with analogous things, images out of those things that seemed appropriate. {As *Qian* is hard and *Kun* is soft, so each thing has its substantial character. This is why the text says: "drawing comparisons to them with analogous things. "} This is why these are called "images." The sages had the means to perceive the activities taking place in the world, and, observing how things come together and go smoothly, they thus enacted statutes and rituals accordingly.

## —3—
# Goldin, Paul R. In Mair, Victor H., Steinhardt, Nancy S., and Goldin, Paul R., eds. 2005. *Hawai'i Reader in Traditional Chinese Culture*. Honolulu: University of Hawai'i Press, 165–168.

These excerpts are from several texts that present the Chinese *wu xing* (five phases) correlative cosmology in all its elaborate detail. "Like Responds toLike" is from the *Spring and Autumn Annals of Mr. Lu*; "The Vast Plan" is *from Exalted Documents* and its *Grand Commentary.* The final two are from the *Huainanzi.*

[ . . . ]

Whenever an emperor or king is about to flourish, Heaven must first cause an omen to appear to the people below.

At the time of the Yellow Emperor, Heaven first caused great earthworms and mole crickets to appear. The Yellow Emperor said, "Earth *qi* prevails." Since Earth *qi* prevailed, he exalted yellow as his color and modeled his activities after Earth.

When it came to the time of Yu, Heaven first caused grasses and trees to appear throughout autumn and winter without dying. Yu said, "Wood *qi* prevails." Since Wood *qi* prevailed, he exalted green as his color and modeled his activities after Wood.

When it came to the time of Tang, Heaven first caused metal blades to appear growing in the water. Tang said, "Metal *qi* prevails." Since Metal *qi* prevailed, he exalted white as his color and modeled his activities after Metal.

When it came to the time of King Wen, Heaven first caused fire to appear, and red rooks with cinnabar writings in their beaks to gather at the altars of Zhou. King Wen said, "Fire *qi* prevails." Since Fire *qi* prevailed, he exalted red as his color and modeled his activities after Fire.

What will replace Fire is surely Water. Heaven will first make it apparent that Water *qi* prevails; and since Water *qi* will prevail, [the new ruler] will exalt black as his color and model his activities after Water. Water *qi* will reach its limit, and then, without our knowing it, the sequence will come full circle and shift back to Earth.

[ . . . ]

The Five Phases:

The first is called Water. The second is called Fire. The third is called Wood. The fourth is called Metal. The fifth is called Earth.

Water moistens and descends.

*Commentary:* If one diminishes one's ancestral temple, fails to pray to and worship one's forebears, lets sacrifices lapse, and opposes the Heavenly seasons, then Water will not moisten and descend.

Fire blazes and ascends.

*Commentary:* If one abrogates administrative laws, chases away meritorious ministers, kills the Heir Apparent, and makes one's concubine one's primary wife, then Fire will not blaze and ascend.

[ . . . ]

What are the five stars?

The eastern quadrant is Wood. Its god is Tai Hao. His assistant, Goumang, grasps the compass and governs spring. Its spirit is the Year Star, its beast the Blue-Green Dragon, its tone *jue,* its days *jia* and *yi.*

The southern quadrant is Fire. Its god is the Blazing Emperor. His assistant, Red Brightness, grasps the steelyard-beam and governs summer. Its spirit is Yinghuo, its beast the Vermilion Bird, its tone *zhi,* its days *bing* and *ding.*

The center is Earth. Its god is the Yellow Emperor. His assistant, the Goddess of the Soil, grasps the marking-line and regulates the four quadrants. Its spirit is the Apotropaic Star, its beast the Yellow Dragon, its tone *gong,* its days *wu* and *ji.*

The western quadrant is Metal. Its god is Shao Hao. His assistant, Rushou, grasps the T-square and governs autumn. Its spirit is the Grand Whiteness, its beast the White Tiger, its tone *shang,* its days *geng* and *xin.*

The northern quadrant is Water. Its god is Zhuanxu. His assistant, Mysterious Obscurity, grasps the steelyard-weight and governs winter. Its spirit is the Morning Star, its beast the Mysterious Warrior, its tone *yu,* its days *ren* and *gui.*

[ . . . ]

Wood vanquishes Earth; Earth vanquishes Water; Water vanquishes Fire; Fire vanquishes Metal; Metal vanquishes Wood. Thus grain is born in the spring and dies in the autumn. Pulse is born in the summer and dies in the winter. Wheat is born in the autumn and dies in the summer. Shepherd's-purse is born in the winter and dies in midsummer.

When Wood is robust, Water is aging, Fire is born, Metal is imprisoned, and Earth is dead. When Fire is robust, Wood is aging, Earth is born, Water is imprisoned, and Metal is dead. When Earth is robust, Fire is aging, Metal is born, Wood is imprisoned, and Water is dead. When Metal is robust, Earth is aging, Water is born, Fire is imprisoned, and Wood is dead. When Water is robust, Metal is aging, Wood *is* born, Earth is imprisoned, and Fire is dead.

—4—
# Knoblock, John, Riegel, Jeffrey. 2000. *The Annals of Lu Buwei: A Complete Translation and Study.* Stanford, CA: Stanford University Press, 60–64.

This work is not to be confused with the older Confucian classic *Chun-qiu, (Spring and Autumn Annals)* of the first period of the Eastern Zhou.

The title of both indicates the importance to the Chinese of using the seasons as the basis for organizing society. Lu Buwei intended this work, composed by a large group of scholars in 239 B.C.E., to include all the knowledge relevant to rulership. Lu was associated with the rise to power of the totalitarian Qin emperor who unified China. The first emperor was despised by the Confucians for his infamous book-burning- and also because they were kept out of office. Lu was even rumored, almost certainly falsely, to be the first emperor's natural father. Imperial politics were labyrinthine; Lu eventually fell out of favor and committed suicide fearing that he would otherwise be executed. For Chinese intellectuals, association with the imperial court was always hazardous, a theme emphasized by Zhuangzi, many of whose anecdotes describe people refusing high office.

Lu's work is of great value because it gives a thorough picture of Chinese thought on the threshold of the imperial period. The selection here gives an example of the rigid protocol prescribed for rulers.

[ . . . ]

A. During the first month of spring the sun is located in Encampment.

At dusk the constellation Triad culminates, and at dawn the constellation Tail culminates.

B. The correlates of this month are the days *jia* and *yi*, the Sovereign Taihao, his assisting spirit Goumang, creatures that are scaly, the musical note *jue*, the pitch-standard named Great Budding, the number eight, tastes that are sour, smells that are rank, and the offering at the door. At sacrifice, the spleen is given the preeminent position.

C. The east wind melts the ice, dormant creatures first begin to stir, fish push up against the ice, otters sacrifice fish, and migrating geese head north.

[ . . . ]

In this month occurs the period "Establishing Spring." Three days before the ceremony marking the Establishing Spring, the grand historiographer informs the Son of Heaven, saying: "On such-and-such a day begins Establishing Spring. The Power that is flourishing is Wood." The Son of Heaven then begins his fast. On the day beginning "Establishing Spring," the Son of Heaven personally leads the Three Dukes, the Nine Ministers, the feudal lords, and the grand officers in welcoming spring at the eastern suburban altar. On returning, he rewards the dukes, ministers, feudal lords, and grand officers in the court. He mandates that his assistants should make known the moral authority of his government, propagate his ordinances of instruction, execute celebratory commemorations, and bestow favors so that they reach down even to the millions of his subjects.

[ . . . ]

The king distributes the tasks of agriculture and commands that field inspectors lodge at the eastern suburban altar. They are to insure that everyone keeps boundaries and borders in good repair and that care is taken as to the straightness of the small pathways between fields. They are skillfully to survey the mounds, slopes, ravines, plains, and marshes to determine which have soil and landforms suitable to grow each of the five grains. In all this they must instruct the people and personally participate in the work.

When before tasks in the fields are announced,
The boundaries have all been fixed,
The farmers will harbor no suspicions.

[ . . . ]

If in the first month of spring the ordinances for summer are put into effect, then winds and rains will not be seasonable, grasses and trees will wither early, and the state will thereupon become alarmed. If the ordinances for autumn are put into effect, the people will suffer a great plague, severe winds and violent rains will frequently occur, and briars, darnel, brambles, and artemisia weeds will flourish together with the crops. If the ordinances for winter are put into effect, floods and heavy rains will cause ruin, frost and snow will do great damage, and the first-sown crops will not mature so that they can be harvested.

—5—
# Xu, Gan. 2002. *Balanced Discourses: A Bilingual Edition.* New Haven, CT: Yale University Press, and Beijing, China: Foreign Languages Press, 137–143; 175–185.

Xu Gan (170–217) lived at the end of the Eastern Han. Like *The Spring and Autumn Annals* of Lu Buwei, the *Balanced Discourses* is an effort to record all important knowledge. The first selection is a comment on loss of ethical sensibility, always an issue in China when ambitious politicians pretended to virtues they did not possess. What is of particular interest in Xu Gan's discussion is his use of a blocked artery as an analogy for hidden moral failings. As so often in Chinese writings, natural phenomena are mainly of interest as similes for social ones.

The second passage exemplifies the Chinese assumption that inaccuracies in the calendar and in astronomical calculation were causes of social disorder, a major concern at the time of writing because the Han dynasty was in terminal decline. The discussion of astronomy and the calendar emphasizes Xu Gan's belief, widely shared throughout Chinese history, that inaccuracies in the calendar resulted in disorder in the human realm.

[ ... ]

Confucius died several hundred years ago. In the interim, no more sages have appeared, the laws of Tang and Yu have become effete, the teachings of the Three Dynasties have waned, the great way has fallen into decline and the point of balance for ordered human relationships has remained in an unsettled state. Thus charlatans and seekers of undeserved fame, taking advantage of the people's long departure from the sagely teachings, have spawned aberrant heresies and created heterodox practices. They use the instructions bequeathed by the former kings to avail themselves of a facade. While externally they conform, in actuality they contravene those instructions; while in appearance they accord, yet their real condition places them at a great remove. They boast that they have attained the truth of the sages, and each of them deploys two-sided explanations and uneven arguments. They have deceived a whole era of people, enticing them with their falsely earned reputations and frightening them with fabricated slander. They cause people to become agitated about success and failure, and in their despondency, they become restless. Having lost their original natures, people are not even aware that they have fallen under their spell, and so, joining with one another, they all adopt the teachings of these men as their model, revering them and drawing close to them. The effect of these heterodox practices upon these people is similar to the effect of a blocked artery. Because the body does not suffer any painful itching or vexations, and because the faculties will function as if still keen, one does not sense that the illness is already acute. When, the time comes, however, the blood and vital energy will suddenly cease flowing. Hence a blocked artery, once developed, will lead to premature death, and this is what those people afflicted by it detest most of all. Being such a malady, not even the physician of Lu could diagnose it nor could Bian Que treat it.

[ ... ]

Long ago, when the sage kings created astronomical systems, they studied the motions of the sun, moon, and stars, and watched the turning of the Northern Dipper; they traced the alternation of the paths of stars and constellations as they crossed the observer's meridian due south; and they understood the significance of the length of the shadows cast by the sun. Thereupon they made instruments to improve the accuracy of their observations, erected gnomons to measure, set up clepsydras to check the timing of star transits, and laid out counters to plot the lengths of the sun's shadow. Only then would the beginning of the year be arranged at the outset, the medial *qi* periods and the nodal *qi* periods be placed in proper order for the rest of the year, the heat and the cold follow one another in orderly succession, and the four seasons proceed without irregularity. Astronomical systems were used by the former kings to promulgate the different periods for the taking of life and the nurturing of life, and, by edict, to set the seasons for conducting various activities, so ensuring that the people of all states would not fail to pursue their occupations.

In the past, "with the decline of Shaohao, the Nine Li tribes threw virtue into disorder, the affairs of the people and the spirits became entangled with one another, and things could not be distinguished. Zhuan Xu acceded to the throne and ordered Zhong, Chief of the South, to be responsible for affairs relating to heaven, and entrusted him with matters concerning making offerings to the spirits. He also ordered Li, Chief of the North, to be responsible for affairs relating to earth, and entrusted him with matters relating to putting the people in order. This was to ensure that the spirits and the people would resume their former regular practices and cease violating one another's domain and being contemptuous of one another. After this, the Three Miao tribes once again acted in the manner of the Nine Li tribes. Yao once more nurtured the descendants of Zhong and Li. Those among them who had not forgotten the old skills were made to restore the canons and instruct the others in their use." Hence the *Book of Documents* says: "Then Yao commanded Xi and He to comply reverently with august heaven and its successive phenomena, with the sun, the moon, and stellar markers, and thus respectfully to bestow the seasons on the people." Thereupon the *yin* and *yang* forces became harmonized, calamities and pestilence did not arise, propitious omens arrived at due intervals, fine crops grew profusely, the people were happy and well, and the ghosts and spirits bestowed blessings. When Shun and Yu received the mandate to rule, they continued the practice [of employing the descendents of Zhong and Li] and so made no errors.

[...]

Hence when the charismatic virtue of the Zhou had declined, the one hundred standards of measurement were abandoned, and the astronomical system lost its standard of reckoning. For this reason, in the first year of the reign of Duke Wen of Lu [626 BCE], there was an intercalated third month. The *Spring and Autumn Annals* criticized him, and its tradition says: "This goes against propriety. In regulating the seasonal ordinances, the former kings made the beginning [the winter solstice] the starting point, determined the correct periods for demarcating the seasonal divisions based on the equinoxes and solstices, and reserved the surplus day, for the year's end. Because they made the beginning as the proper starting point, no error was made in the proper order of the seasons; because they determined the correct periods for demarcating the seasonal divisions, the people were not confused; and because they reserved the surplus days for the year's end, affairs proceeded without any error."

[ . . . ]

Emperor Wu [r. 141–87 BCE] restored the standards of the former kings and followed the old regulations. He summoned scholars of the Five Classics and men skilled in the specialist arts and numerology to deliberate upon and fix the Han calendrical system, and also changed to using the revised astronomical system developed by Deng Ping, [Luoxia Hong, and others].The epoch was calculated from the beginning of the Grand Inception reign period [24 December 105 BCE]. Thereafter, the equinoxes, solstices and the commencement and conclusion of the four seasons maintained a regular sequence, and the first and last quarters of the moon, the full moon, the last day of the moon, and the new moon could be accurately predicted.

[ . . . ]

Thereupon Chief Commandant Liu Hong of Guiji [alt. Kuaiji] developed the Celestial Images astronomical system so as to trace the movements of the sun, moon, and the stellar markers. To this day, the precision of his system is verified by observation of the heavenly bodies and their movements. With Emperor Ling's recent passing, the capital has been thrown into a state of utter chaos and, unfortunately, matters of astronomy have been abandoned!

The matter of astronomical systems from former generations until the present having thus been reviewed, it is evident that when emperors and kings rose to power, they all respectfully employed natural time to assist them in conducting the affairs of man. Thus when Confucius wrote the *Spring and Autumn Annals* and recorded the affairs of man, he did so with reference to natural time so as to make it clear that the realms of heaven and man are integrated through codependence. Accordingly, if a ruler failed to observe the equinoxes, solstices, and the commencement and conclusion of the four seasons, then Confucius would not record the particular season or the month in which such matters occurred. This was done so as to criticize such rulers for being lax and remiss.

Astronomical systems are the means by which the sage fathoms the profundity of the sun and moon and thereby thoroughly penetrates the genuine condition of the mysterious and the sublime. If the most precise astronomical systems are not employed, then who would be able to apply their thoughts to these matters? Today, in very general terms, I have discussed the old systems of several experts and put them together in this essay. I hope that it will provide a record of what has been omitted from and what has been preserved of the principles of astronomical systems, for the benefit of enlightened men of future times.

—6—
# Smith, Brian K. 1999. *Classifying the Universe: The Ancient Indian Varna System and the Origins of Caste*. Oxford: Oxford University Press, 136–137.

This is an excerpt from a version of the *Satapatha Brahmana,* an early text somewhat later than the *Vedas,* describing part of the process of creation. Notable is the emphasis on how the gods organized the space of the earth, largely to accord with compass directions. Except for the specific mention of gods, this text is reminiscent of Chinese cosmogonic texts. The concern that space be ordered is of interest. We take the physical organization of the earth for granted; it was established long before we were born. However, when the human species arose, there was no such structure to the world and in order for society to exist, it had to create a spatial order for itself. In Hinduism, the four varnas or castes were each associated with a compass direction so that the shape of society was correlated with that of the physical environment.

[ . . . ]

These five deities [Indra, Soma, Agni, Vāc, and Prajāpati] then sacrificed with that wish-fulfilling sacrifice; whatever wish they sacrificed for was fulfilled. . . . After they had sacrificed they saw the eastern (or "forwards") quarter and made it into the eastern quarter—this is that eastern quarter [as it exists today]. Therefore creatures here move along in a forward direction, for they made that the forward (or "eastern") quarter. "Let us improve it from here," they said. They made it into nourishment *(urj).* "Furthermore, may we see this nourishment" they said, and it became yonder sky. They then saw the southern *(daksina)* quarter and made it the southern quarter—this is that southern quarter. This is why the *daksina* [cows] stand to the south [of the altar] and are driven up from the south, for they made that the southern quarter. "Let us improve it from here," they said. They made it into a world. "Furthermore, may we see this space," they said, and it became the atmosphere, for that world is the atmosphere.

Just as the foundation here in this world is obviously the earth, so the foundation there in yonder world is clearly this atmosphere. And because, being here on earth, one does not see that world, people say, "That world is invisible." They then saw the western *(pratici)* quarter, and made it into expectation *asa* [which can also mean "quarter" or "region"]. Thus, having moved forwards [or to the east] one procures [what he desires], then he goes [back] to that [western] quarter. For they made that [quarter] into expectation. "Let us improve it from here," they said. They made it into excellence *(sriyai)*. "Furthermore, may we see this excellence," they said, and it became this earth, for this [earth] is excellence. Thus, one who gains the most from this [earth] becomes superior *(srestha)*. They then saw the northern *(udici)* quarter, and made it into the waters. "Let us improve it from here," they said. They made it into proper order *(dharma)*, for the waters are proper order. Thus, whenever water comes to this world everything is in accordance with order.

## —7—
## Danielou, Alain. 1964. *The Gods of India: Hindu Polytheism*. Trans. from the French. Reprinted 1985, New York: Inner Traditions International, 30–34; 222–225.

The translator, Alain Danielou, lived in India for many years as a Hindu and thus came to understand India's traditional culture from within. Although this passage is a translation of a translation, Danielou's vivid style is preserved and brings across the intellectual appeal of Indian metaphysics more clearly than most literal versions. Clarity is attained by combining sections with related themes from disparate texts. Though it gives less of the feeling of the actual texts, which tend to be quite prolix with devotional elements frequently inserted, it does provide a good entry into the complexities of Indian cosmology and metaphysics.

### Illusion and Ignorance

*Maya* is equally the source of the cosmos and of the consciousnesses that perceive it. Both are interdependent. The nonperceived cosmos has no existence and the nonperceiving consciousness no reality.

Manifestation exists only in relation to perception. If none perceived the cosmos, one could not say that it exists. Hence the principles of the senses, like the principles of the elements, are considered the causes of manifestation, the forms of the Creator.

The perceiving consciousness is the necessary corollary of the manifest-ing power; the living being and the divine Being exist only in relation to one another. Hence the necessity of the individual consciousnesses of the living beings for the creating power, the dependence of the Creator on his creation.

In the microcosm, that is, the manifested individuality, the power-of-illusion *(maya)* becomes the power-of-ignorance *(a-vidya)* or un-knowing *(a-jnana)*, the perceiver of the forms of illusion.

"Whether through knowing or through unknowing, all things take their reality from that which perceives them" (Karapatri, "Sri Bhagavati tattva," p. 242).

The difference between the power-of-illusion *(maya)* and the power-of ignorance *(avidya)* is that through *maya*, through the cosmic illusion, the manifestation of the universe takes place, the Abysmal Immensity manifests itself in the immense cosmos, while through ignorance centers of perception are formed so that the illusion may be perceived and thus become a relative reality. It is through ignorance *(avidya)* that individual beings come into existence as distinct entities.

Here unknowing *(ajnana)* is not mere absence of knowledge but rep-resents a state beyond knowledge, the nature of which is transcendent Being.

"Illusion is the experience of the form of darkness (or unknowing) [by which the substratum is veiled]." *(Nrsimha-uttara-tapini Upanisad* 9.4. [40])

"Perception veiled by unknowing" is the intrinsic nature of the universe. This is only possible because unknowing is the nature of existence; the nonexistent cannot veil. Unknowing, like illusion, is a veil.

[ ... ]

Unknowing is the source of experience. If unknowing were mere absence such experience could not take place, for all illusion is based on conscious-ness.

[ ... ]

"Nature *(prakrti)* is that from which [things] are born." *(Panini Sutra* 1.4.30.)

"Nature is that which acts constantly. It is the first basis, the state of balance of *sattva, rajas,* and *tamas." (Sankhya-tattva-kaumudi,* commentary on *Sankhya Karika* 3. [44])

"Being the Nature *(prakrti)* [of things] implies being the immanent cause [of all things] through the absolute hierarchy [of causes and effects]. It is defined as transcendent action, that is, action in the form of evolution. Nature *(prakrti)* is often taken as the synonym of Energy *(sakti)*, of the Unborn *(aja)*, of the First-Basis *(pradhana)*, of the Nonevolved *(avyakta)*, of Disintegration *(tamas)*, of Illusion *(mayli)*, and of Ignorance *(avidya)*." (Vijñana Bhiksu, *Sankhya-sara* 1.3. [45])

Creation is born of the power of knowing, more or less veiled by that of action. The veil is thickest in inanimate matter and gradually lighter in plants and mountains, insects, birds, wild and tame animals, in man, in angels and genii, and in gods.

[...]

"She whose shape even the Creator and the other gods cannot know is called 'the unknowable' *(ajneya)*. She whose end cannot be found is called 'the endless' *(ananta)*. She who alone is everywhere present is called 'the One' *(eka)*.

"In all knowledge she is the transcendent consciousness; in all voids she is the Void. She, beyond whom there is no beyond, is sung as Beyond-Reach (Durga)." *(Devf Upanishad* 26–27. [47])

[...]

Everything in the work of manifestation is intended to create the illusion and to prevent the realization of the basic oneness of all beings, this would lead to the destruction of the notion of I-ness, which is the power of cohesion that holds together the individual being, the witness that gives to the cosmos. Any weakening of the centripetal tendency characteristic of the individuality is contrary to the process of the world's creation. The aim of any creator is to prevent a realization which would destroy his creation. This is why "the Soul is not within the reach of the weak." It has to be conquered by going against all the forces of Nature, all the laws of creation.

Nature is the expression of the Creator's thought; it is the power that creates forms. It is also the power that prevents escape from the world of forms. In his efforts toward liberation man has to overcome all the artifices of Nature, which ever delude him. All revealed knowledge is intended to keep man a manifest individuality. We can therefore understand why all the rules of social morality, all the forms of worldly and sacred knowledge, all the bonds of religion its rites, seem basically intended to take man away from the path of liberation through the promise of heavenly bliss or pleasures or other advantages.

All our means of perception are oriented outward. Like insects attracted by a flame, we are attracted by the sights and sounds of the external world that are intended to prevent us from looking inward.

[...]

When the yogi, the seeker of liberation, makes some headway in his attempt to free himself from all bonds, Nature offers him new abilities, new temptations, new achievements, to bring him back into her power. Each of the qualities of Nature, instead of being a way toward liberation, becomes an instrument of bondage.

[...]

"Know the power-of-disintegration *(tamas)* to spring from ignorance; it misguides all living beings and binds them, O scion of Bharata! through carelessness, laziness, and sleep." *(Bhagavadgita 14.6–8. [51])*

[ . . . ]

## The *Linga* Is the Universe

"From the signless comes forth the sign, the universe. This sign is the object of word and touch, of smell, color, and taste. It is the womb of the elements, subtle and gross." *(Linga Purana 1.3.3–4. [417])*

"Basic-Nature *(pradhana)* is thus called the *linga*. The owner-of-the-*linga* is supreme divinity." *(Linga Purana 1.17.5. [418])*

"The root of the *linga* is the signless. The *linga* itself is Unmanifest-Nature *(avyakta)*. Thus Siva himself is sexless. The *linga* is the thing-of-Siva." *(Linga Purana 1.3. [419])*

"Siva as the undivided causal principle is worshiped in the *linga*. His more manifest aspects are represented in anthropomorphic images. All other deities are part of a multiplicity and are thus worshiped in images." (Karapatri, Sri Siva tattva," *Siddhanta*, II, 1941–42.)

"The Sun is envisaged as the progenitor of the worlds, hence its symbol is that of procreation." *(Siva Purana1.16.105. [420])*

"MANIFEST NATURE *(vyakta)*, the universal energy, is shown as the *yoni*, or female organ, embracing the *linga*. Only under the shape of a *linga*, giver of seed, can Siva be enveloped by the *yoni* and become manifest.

"The symbol of the Supreme-Man *(purusa)*, the formless, the changeless, the all-seeing eye, is the symbol of masculinity, the phallus or *linga*. The symbol of the power that is Nature, generatrix of all that exists, is the female organ, the *yoni*." (Karapatri, "Lingopasana-rahasya," *Siddhanta*, II, 1941–42, 154.)

The *linga* stands in the *yoni*, which is the power of manifestation, "the womb of the real and the unreal." *(Sukla Yajur Veda, Vajasaneya Samhita 13.3; Taittiriya Sarhhita 4.2.8.2. [421])* "The universal womb in which all that is individual ripens." *(Svetasvatara Upanisad 5.5. [422])*

The *linga* fecundates this womb.

"The vast Immensity [the principle of intellect] is the womb in which I [Krsna] deposit my seed. From it are born [Manifest Intellect, the first element] and then [in hierarchy] all the elements [and all the creatures]." *(Bhagavadgita 14.3. [4,23])*

[ . . . ]

IN THE unmanifest state there is a perfect balance. Then all the gods appear as one; there is no perceptible duality, no distinction of a positive and a negative force. As soon as the first tendency toward manifestation appears in the undifferentiated substratum duality is already present. This duality has the character of complementary poles of attraction, a positive and a negative tendency, which will be manifested in the whole of creation by male and female characteristics, There can be no creation without the relation of opposites. There could be no creation from Siva alone or from Nature *(prakrti)* alone. The union of a perceiver and a perceived, of an enjoyer and an enjoyed, of a passive and an active principle, of a male and a female organ, is essential for creation to take place. The union of Supreme Man and Nature is represented in the copulation *(maithuna)* of the lord-of-sleep (Siva) and Faithfulness (Sati), that is, Energy (Sakti).

Transcendent manhood is the immanent cause of creation; transcendent womanhood, the efficient cause. In the microcosm these principles are mainly apparent in the sex organs, which represent the most essential physical function of all beings.

—8—

# Needham, Joseph with the collaboration of Wang Ling. 1959. *Mathematics and the Science of the Heavens and the Earth*, Vol. 3. New York: Cambridge University Press, 284–286; 367–369.

The first portion consists of descriptions of the gnomon in two ancient texts, the *Zouzhuan* commentary to the *Chunqiu* and the *Zhouli*. Though more than two thousand five hundred years old, these texts describe correct methods to determine solstices and equinoxes.

The second excerpt is by the famous Jesuit missionary Matteo Ricci, who lived in China in the early seventeenth century and was the possibly the first Westerner to fully learn both the spoken and written languages. A Chinese observatory is described and, though the tone is condescending, as were most writings of early Europeans resident in Asia, it is evident that Chinese astronomers had attained a high level of technical expertise.

[ ... ]

In the fifth year of Duke Hsi, in spring, in the first month (December), on a *hsin-hai* day, the first of the month, the sun (reached its) furthest south point. The duke (of Lu), having caused the new moon to be announced in the ancestral temple, ascended to the observation tower *(kuan thai)* in order to view (the shadow), and (the astronomers) noted down (its length) according to custom. At every equinox and every solstice, whether at the beginning *(chhi)* of spring or summer, or at the beginning *(pi)* of autumn or winter, it is necessary to note the appearance of the clouds and vapours, for (prognostication and) preparations against coming events.

[ ... ]

The Ta Ssu Thu (a high official) [says the *Chou Li*], using the gnomon shadow template, determines the distance of the earth below the sun, fixes the exact (length of the) sun's shadow, and thus finds the centre of the earth.... The centre of the earth is (that place where) the sun's shadow at the summer solstice is 1 ft. 5 in.

[...]

Not only in Peking, but in this capital also (Nanking) there is a College of Chinese Mathematicians, and this one certainly is more distinguished by the vastness of its buildings than by the skill of its astronomers. They have little talent and less learning, and do nothing beyond the preparation of almanacs on the rules of calculation made by the ancients—and when it chances that events do not agree with their computations they assert that what they have computed is the regular course of things, and that the aberrant conduct of the stars is a prognostic from heaven about something which is going to happen on earth. This something they make out according to their fancy, and so spread a veil over their blunders. These gentlemen did not much trust Fr. Matteo, fearing no doubt lest he should put them to shame; but when at last they were freed from this apprehension they came and amicably visited him in the hope of learning something from him. And when he went to return their visit he saw something that really was new and beyond his expectation.

There is a high hill at one side of the city, but still within the walls. On the top there is an ample terrace, capitally adapted for astronomical observation, and surrounded by magnificent buildings erected of old. Here some of the astronomers take their stand every night to observe whatever may appear in the heavens, whether meteoric fires or comets, and to report them in detail to the emperor. The instruments proved to be all cast of bronze, very carefully worked and gallantly ornamented, so large and elegant that the Father had seen none better in Europe. There they had stood firm against the weather for nearly two hundred and fifty years, and neither rain nor snow had spoiled them.

There were four chief instruments. The first was a globe, having all the parallels and meridians marked out degree by degree, and rather large in size, for three men with outstretched arms could hardly have encircled it. It was set into a great cube of bronze which served as its pedestal, and in this box there was a little door through which one could enter to manipulate the works. There was, however, nothing engraved on the globe, neither stars nor terrestrial features. It therefore seemed to be an unfinished work, unless perhaps it had been left that way so that it might serve either as a celestial or a terrestrial globe.

The second instrument was a great (armillary) sphere, not less in diameter than that measure of the outstretched arms which is commonly called a geometric pace. It had a horizon (-circle) and poles; instead of (solid) circles it was provided with certain double hoops (armillae), the void space between the pair serving the same purpose as the circles of our spheres.

All these were divided into 365 degrees and some odd minutes. There was no globe representing the earth in the centre, but there was a certain tube, bored like a gun-barrel, which could readily be turned about and placed at any degree or altitude so as to observe any particular star, just as we do with our vane sights—not at all a despicable device.

The third piece of apparatus was a gnomon, the height of which was twice the diameter of the former instrument, erected on a rather long slab of marble which pointed to the north. This had a channel cut round the margin, to be filled with water in order to determine whether the slab was level or not, and the style was set vertical as in hour-dials. We may suppose this gnomon to have been erected that by its aid the shadow at the solstices and equinoxes might be precisely noted, for both the slab and the style were graduated.

The fourth and last instrument, and the largest of all, was one consisting of, as it were, 3 or 4 huge astrolabes in juxtaposition; each of them having a diameter of such a geometrical pace as I have specified. The fiducial line, or Alhidada, as it is called, was not lacking, nor yet the Dioptra. Of these astrolabes, one having a tilted position in the direction of the south represented the equator; a second, which stood crosswise on the first, in a north and south plane, the Father took for a meridian, but it could be turned round on its axis; a third stood in the meridian plane with its axis perpendicular, and seemed to represent a vertical circle, but this also could be turned round to show any vertical whatever. Moreover, all these were graduated and the degrees marked by prominent (metal) studs, so that in the night the graduation could be read by touch without any light. This whole compound astrolabe instrument was erected on a marble platform with channels round it for levelling.

# —9—
# Unschuld, Paul. 1986. *Medicine in China: A History of Pharmaceutics*. Berkeley: University of California Press, 98–99.

This passage from the *Bencao Yanyi* by the Song physician Kuo Zongshi is a fairly typical example of Chinese medical theory based on correlative metaphysics, or as termed by Unschuld, systematic correspondence. Here we find yin and yang and heaven, earth, and man. As true in Indian traditional medicine as well, there are observations of individual differences that must be considered in selecting prescriptions, still an issue in scientific medicine today. Kuo also shows awareness of what we would now term psychosomatic or mind-body medicine in that he observes that happiness or sadness affect health. The work is thoughtful but the empirical element is limited by the closed metaphysical system of interpretation.

[ ... ]

The life of men is truly founded on a concentration of *yin* and *yang* influences. If one cannot balance the *yin* and *yang* influences, he will do harm to his life. Therefore, the treatise *Pao-ming ch'uan-hsing p'ien* comments: "Man emerges through the influence of heaven and earth." It further states: "Heaven and earth allow their influences to combine, and man emerges." That is possible because [what is categorized as] *yang* transforms itself into finest influences, and [what is categorized as] *yin* materializes itself into a shape. The transformation of the ethereal soul corresponds to the change of *yang* [aspects] into finest influences. The transformation of finest matter influences into items corresponds to the materialization of *yin* [aspects]. When the *yin* and *yang* influences unite, the spirit will be in their midst. Therefore the treatise *Yin-yang ying-hsiang ta-lun* says: "Movement and motionlessness of heaven and earth are regulated by the spirit-brilliance." This shows that the spirit-brilliance cannot be grasped with [the categories of] *yin* and *yang*. That corresponds to the statement in the *I[-ching]*: "That which cannot be measured by [the categories of] *yin* or *yang* is called spirit." Therefore it is said: The spirit must not be taxed excessively, otherwise it will be exhausted. The physical form must not be burdened excessively, otherwise it perishes. Consequently, finest essence, finest influences, and the spirit are the most important foundations of human [existence]. One must carefully maintain these foundations. Those who possess knowledge nourish their spirit and economize on their influences in order to strengthen their foundations.

[ ... ]

There are rich and poor, young and old people, and their respective illnesses must be considered separately. There are illnesses that have just begun and those that have lasted a long time, as well as [cases of] depletion and repletion, and their respective etiologies, which require different drugs. The hearts of the people, like their faces, differ from individual to individual. Not only are the hearts different but so are the depots and the palaces of the body. And if the depots and the palaces are different [from person to person], how would one manage to cure one [specific] illness in a large number of people with one and the same drug? Therefore, Chang Chung-ching said: There is a difference between high locations and low-lying regions. The nature of things may be soft or hard. There are differences in eating and living. For this reason, Huang-ti posed the question about the four cardinal points, and Ch'i Po mentioned the four therapeutic possibilities. In order to attain successes against illnesses, one must take both of these into consideration.

Consequently, it is crude negligence to combine drugs according to [fixed] prescriptions and then to use them on a broad basis. Wealthy people, for example, appear outwardly happy, but their minds suffer. If clothes and food are available in sufficient amounts, one seems outwardly content. But if, at the same time, the thoughts are burdened by too many reflections, one's mind suffers. Ch'i Po said: The illnesses originate in the vessels. If someone is outwardly happy, there is external repletion. If someone's mind suffers, there is an inner depletion. Therefore, the illnesses originate in the vessels. The lives [of the rich and] of the poor are marked by different nourishment. Sadness, joy, and thoughts are also different. In every case, treatment must be oriented toward the individual person. The medical men of later times really did not pay any attention to this rule; they cut it off and did not follow it. The loss suffered in this manner is considerable!

# —10—
# Furth, Charlotte, Zeitlin, Judith T., Hsiung, Ping-chen. 2007. *Thinking with Cases: Specialist Knowledge in Chinese Cultural History*. Honolulu: University of Hawai'i Press, 154–155.

This excerpt is by the Song dynasty physician Qian Yi (1031–1113) and clearly shows that he recognized that boiling water would prevent diarrhea in small children, still a major cause of death in the developing world. He also recognized that purging (use of laxatives) would make what we can now recognize as dehydration worse. Whether the herbal medicines contributed to the recovery of the infant cannot be determined, but the empirically based prescription of hydration clearly did. There is no evidence, however, that his life-saving realization was widely transmitted.

[ . . . ]

The five-year-old son of an official, Zhu, was feverish at night. When the boy awoke in the morning, he was normal. Various common practitioners treated it either as a case of "damage from cold"; or as a case of a "hot" disorder. They prescribed cooling drugs for the child to disperse the heat, but the child was not cured. His symptoms were excessive mucus and drowsiness. Other doctors used Powdered Iron Pills [*tiefen yuan*] to purge the mucus, but the child's illness worsened. On the fifth day he had a great craving for drink. Qian said, "One must not purge." Then he took one *liang* [fifty grams] of finely chopped White Atractylodes Rhizome [*baishu san*], boiled it until there were three *sheng* [liters] of juice, gave it to the child and let him drink as much as he desired. Zhu asked, "If he drinks a lot, won't he have diarrhea?" Qian replied, "Without un-boiled water one develop diarrhea. Even if the child develops diarrhea, it should not be anything strange. But one must not purge." Zhu again asked, "What disorder should be treated first?" Qian replied, "[My medication is] to quench thirst, stop phlegm, reduce fever, and clean the insides all with this one prescription." When evening came, the prescribed dosage had been used up. Qian again examined the situation and ordered, "He can take yet three more *sheng*." Three more *sheng* of Baishu Powder Drink were prepared and administered, whereupon there was some improvement. On the third day, [the patient received] yet another three *sheng* of the Baishu Powder Drink and the child was no longer thirsty, and also free of mucus. [Qian then] administered two doses of Donkey Glue Powder *[ajiao san]* and [the child] recovered.

# —11—
# McKnight, Brian E., trans. 1981. Sung Ts'u *The Washing Away of Wrong. Science, Medicine, & Technology in East Asia 1*. Ann Arbor: Center for Chinese Studies, The University of Michigan, 106–114.

This passage is a striking example of empiricism in Chinese medicine, in this case forensic pathology. While correlative elements are present, emphasis is on actual observation of the corpse and the surrounding crime scene. It shows the unflinching attitude necessary for medicine to advance. Criminal justice was taken very seriously in traditional China; punishment was harsh. It is possible that the importance of forensic medicine for government favored reliance on systematic observation of evidence rather than on inference from metaphysical theories.

[ … ]

When people have committed suicide by hanging, the eyes will be closed, the lips and mouth black, and the mouth open with the teeth showing. If hanged above the Adam's apple, the mouth will be closed, the teeth firmly set, and the tongue pressed against the teeth but not protruding. *(It is also said that the teeth will be slightly biting the tongue.)* If hanged below the Adam's apple, the mouth will be open and the tip of the tongue will be sticking out past the teeth two- to three-tenths of an inch. The face will be purple red in color, at the two corners of the mouth and on the chest there will be frothy saliva, the two hands will be clenched over the thumbs, and the toes of the two feet will be pointing downwards. On the legs there will be bloody marks resembling the burns of moxibustion.

The stomach and lower abdomen will both be pendulous and blue black in color. The victim will have excreted both urine and feces. There may be a few drops of blood at the anus. The scar on the throat will be purple and red, or a muddy black. It will cross behind the head from the left ear to the right ear at the level of the hairline and have a horizontal length of nine inches to one foot. *(One report says that in* an *adult male, it will measure one foot one inch, and in* a *woman, one foot.)* If the victim's feet are off the ground, the scar on the throat will be deep. If not, then it will be shallow. If the victim was fat, the scar will be deep. If thin, then shallow. If a thin hemp cord or fiber rope was used, and the victim committed suicide by hanging himself from some high place until he strangled, then the mark will be deep. If the victim used a strip of silk cloth or a dressed silk scarf or some such material, or if the victim was in some lower place, then the scar will be less deep. When someone commits suicide by hanging in a low place, the body will often be recumbent either on its side or on its face. If lying on its side, then the mark will slant up across the base of the throat. If prone, then the scar will be vertical, rising from the base of the throat to the ear. Often it will not reach to the hairline at the back of the head.

[ . . . ]

There are often cases where suicide by hanging has been committed in a home by maidservants, manservants, or other nonrelatives living in a household. The people of the household, being ignorant of the law and seeking to avoid the smell and evade the inquest, move the corpse outside and hang it up again. In such cases, because of the shifting of the relative position of the old scar, there will be two scars. The old scar will be purple red with traces of blood under the skin. The scar from the body's having been hung up after being moved will be white with no traces of blood. It will be clear that the corpse has been moved.

## —12—
# Furth, Charlotte. In Mair, Victor H., Steinhardt, Nancy S., and Goldin, Paul R., eds. 2005. *Hawai'i Reader in Traditional Chinese Culture.* Honolulu: University of Hawai'i Press, 440–443.

This passage is by Zhu Zhengheng (1281–1358), a famous literati physician of the Song and Yuan dynasties, and gives a vivid sense of the nature of medical practice at the time. Observation is emphasized but interpretation of what is observed refers back to correlative metaphysics. That only those with sturdy constitutions should be treated with purging suggests awareness of the debilitating effects of such treatment, yet the mode of treatment itself is not questioned. The macro- and microcosmic parallel is apparent in the analogy between medical therapy and how evil rulers were overcome.

## Learning from a Teacher

In the summer of 1325 I first heard of Luo Taiwu [Luo Zhiti] from Chen Zhiyan [a poet]. I proceeded to pay my respects as a pupil. He rejected me with curses a half a dozen times. Only after three months of repeated visits did he finally begin to receive me. In this way I saw Master Luo treat a disorder in a monk who was suffering from emaciation, exhaustion, and jaundice. Master Luo inquired as to the origins of the disorder. The sufferer was from Sichuan, and at the time when he left to take religious vows his mother was still at home. After traveling to Zhejiang as a devotee of the Buddha for seven years, suddenly one day he thought of his mother and could not repress his longing to return home. His money belt was empty, so he simply faced west night and day, weeping. This made him sick. At the time the monk was twenty-four years old. Luo ordered that he rest peacefully at night in an adjoining chamber, and every day he gave him beef, pig belly, and sweet fat simmered and mashed into a pulp. This went on for more than two weeks. In addition, from time to time he offered soothing instructions and encouraging words. Further he said, "I will give you a money order worth ten silver pieces for travel expenses. I do not want repayment; I only wish to save your life." He watched until the young monk's physical form was somewhat revived; then he gave him a one-day treatment of three packets of qi-leading infusion with peach kernel" [to purge his bowels]. The patient evacuated bloody clots and phlegm accumulations. When this was finished, the next day Luo just gave him well-cooked vegetables and rice porridge and nothing more. After another two weeks the young man was his former self, and after two more weeks he was given his ten silver money order and went on his way.

This made me really understand "attack" [purging] therapies. If you pre-scribe to cleanse the bowels, the patient should be in a condition of sturdy repletion, and have a sound constitution. Otherwise, as pathogens leave the body, its healthy, orthodox qi will also be damaged; small disorders will become big ones and big ones will kill.

[ ... ]

Depletion disorders and disorders of phlegm may seem like spirit pos-session. . . . A robust matron, Mistress Jin, attended a feast in hot summer weather. On her return her mother-in-law made inquiries and concluded that she had taken an improper seat at the banquet table. As a consequence the woman was intensely ashamed and this illness resulted. Her speech became incoherent and included the phrase "Th-This slave is at fault!" repeated over and over. Her pulses were all rapid and strung. I said, "This is not a disorder from spirit pathogens. If you will only replenish her spleen, cool down her heat, and lead her phlegm out for a few days, she will be well." Her family did not believe me and called several shamans who spurted holy water and chanted incantations. In a little more than a fortnight she was dead.

   Some people said, "The disorder was not due to an evil spirit, yet it was treated as if it was. What in fact so suddenly led to her death?" I said, "She went to a banquet in summer, when the weather outside was steamy hot. Spicy food went in her mouth, and her internal landscape became stagnant and hot. This was made worse by long-standing accumulations of phlegm, to which were added her shame and dejection. Suffering from phlegm and heat, what worse could be said? The shamans' method of frightening the spirits with magic rods alarmed her psyche, and her blood was disturbed; spurting holy water covered her skin and closed her pores so sweat could not flow. When sweat could not flow, steam heat within enflamed her; her blood could not be tranquilized, so that yin evaporated and yang could not survive alone. Of course she died!"

# —13—
# Zysk, Kenneth. 1991. *Asceticism and Healing in Ancient India: Medicine in the Buddhist Monastery.* Oxford: Oxford University Press, 22–25.

The first excerpt is from the *Taittiriya Samhita* (late Vedic period, ca. 900–500 B.C.E.) and shows the disdain of the Brahmin caste for physicians, a not common attitude in the ancient world, and also the extreme emphasis on elaborate rituals to maintain purity.

   The second excerpt is from a surgical treatise, the *Susruta Samhita* dating from the early centuries of the current era. Medicine and surgery are now elevated to divine origin and therefore are more respectable. The text purports to describe a successful head reattachment, something modern medicine has yet to attempt, but an important element in Indian mythology. It represents adaptation of this earlier myth to enhance the prestige of surgery. The third excerpt is from the *Caraka Samhita*, one of the major texts of classical Indian medicine. We can see in it the continued obeisance to the authority of tradition as well as the close connection of healing to ritual observances to gods.

[ . . . ]

The head of the sacrifice was cut off; the gods spoke to the Asvins: "you two are indeed physicians, [therefore] replace this head of the sacrifice." The two replied: "let us choose a choice [thing]; let now a ladleful [of Soma] be drawn here also for us." They drew this Asvin portion [of Soma] for those two; thereupon, verily, the two replaced the head of the sacrifice; [hence] when the Asvin portion is drawn, [it is] for the restoration of the sacrifice. The gods spoke to those two: "these two physicians, who roam with humans, [are] very impure." Therefore, medicine is not to be practiced by a Brahman, for he, who is a physician *bhisaj*, [is] impure, unfit for the sacrifice. Having purified those two with the Bahispavamana [Strotra], [the gods], drew this Asvin portion for them. Therefore, when the Bahispavamana has been chanted, the Asvin portion is drawn. On account of that, the Bahispavamana should be reverently performed by the one who knows thus; verily the means of purification is the Bahispavamana; indeed, he purifies himself. [The gods] deposited the healing [powers] of those two in three places: a third in fire [Agni], a third in the waters, [and] a third in the Brahman caste. Therefore, having placed the water vessel to one side [and] having sat down to the right of a Brahman, one should practice medicine. To be sure, as much medicine as he practices by this means, his work is accomplished.

[ . . . ]

Indeed, this branch [i.e., *salya]* is first because of its previous [use] in curing wounds from attacks and because of its [use] in the joining of the head of the sacrifice, for, as it is said: "the head of the sacrifice was cutoff by Rudra. Thereupon, the gods approached the Asvin twins [saying,] 'lords, you two will become the best among us; [therefore,] the head of the sacrifice is to be joined by you two lords.' Those two replied, 'let it be so.' Then, for the sake of those two, the gods appeased Indra with a share of the sacrifice. The head of the sacrifice was joined by those two." And also among the eight divisions of *ayurveda*, this *[salya* division] is indeed supposed superior because it does its work quickly, because of its use of blunt and sharp surgical instruments, and cauterization with heat and caustic medicines, and because of its congruence with all other divisions. Therefore, this *[salya* division] is eternal, sacred, heavenly, famous, providing longevity, and also giving a livelihood. Brahma proclaimed [it first], Prajapati [learned it] from him, the Asvin twins from him, Indra from the two Asvins, and I [i.e., Dhanvantari] [learned it] from Indra. Here now it is to be taught by me [i.e., Dhanvantari] to those desiring [it] for the good of humankind.

[ . . . ]

Therefore, by the physician who has inquired about [which Veda an ayurvedic practitioner should follow, (previous verse)], devotion to the *Atharvaveda* is ordered from among the four [Vedas]: *Rgveda, Samaveda, Yajurveda,* and *Atharvaveda.* For it is stated that the sacred knowledge of the fire priests *[atharvans]* is medical science because [it] encompasses giving gifts, invoking blessings, sacrifice to deities, offering oblations, auspicious observances, giving burnt offerings, restraint of the mind, and recitation of magico-religious utterances, and so on; and medical science is taught for the benefit of long life.

# —14—
# Lad, Vasant. 1984. *Ayurveda, The Science of Self-Healing: A Practical Guide.* Wilmot, WI: Lotus Press, 100–103.

This passage is taken from a popular contemporary work on Ayurvedic healing by an Indian Ayurvedic physician who has become established in the United States. In coming West, traditional Asian medicine is modified to suit its new culture, for example, elimination of mercury-based medications in Ayurveda. This excerpt gives a good sense of how Ayurveda combines medical treatments with lifestyle elements, such as massage with vegetable oil and dietary modification.

## SUGGESTIONS FOR A CREATIVE, HEALTHY LIFE
### Routine

- Awaken before sunrise.
- Evacuate bowels and bladder after awakening.
- Bathe every day to create a sense of bodily freshness.
- Twelve *pranayamas* in the morning or evening create freshness of mind and body.
- Do not take breakfast after 8:00 A.M.
- Wash hands before and after eating.
- Brush teeth after meals.
- Fifteen minutes after meals take a short walk.
- Eat in silence with awareness of food.
- Eat slowly.
- Each day massage the gums with the finger and sesame oil.
- Fast one day a week to help reduce toxins in the body.
- Sleep before 10:00 P.M.

### Diet and Digestion

- One teaspoon of grated fresh ginger with a pinch of salt is a good appetizer.

- Drinking *lassi* (buttermilk) with a pinch of ginger or cumin powder helps digestion.
- A teaspoon of *gnee* with rice helps digestion.
- A glass of raw, warm milk with ginger taken at bedtime is nourishing to the body and calms the mind.
- Overeating is unhealthy.
- Drinking water immediately before or after taking food adversely affects digestion.
- Prolonged fasting is unhealthy.
- Consuming excess water may produce obesity.
- Excess intake of cold drinks reduces resistance and creates excess mucus.
- Store water in a copper vessel or put copper pennies in the water. This water is good for the liver and spleen.
- Taking a nap after lunch will increase *kapha* and body weight.

## Physical Hygiene

- If possible, gaze at the rays of the sun at dawn for five minutes daily to improve eyesight.
- Gazing at a steady flame, morning and evening for ten minutes, improves eyesight.
- Do not repress the natural urges of the body, i.e., defecation, urination, coughing, sneezing, yawning, belching and passing gas.
- During a fever, do not eat and observe a ginger tea fast.
- Rubbing the soles of the feet with sesame oil before bedtime produces a calm, quiet sleep.
- Application of oil to the head calms the mind and induces sound sleep.
- Oil massage promotes circulation and relieves excess *vata*.
- Do not sleep on the belly.
- Reading in bed will injure the eyesight.
- Bad breath may indicate constipation, poor digestion, an unhygienic mouth and toxins in the colon.
- Body odor indicates toxins in the system.
- Lying on the back for fifteen minutes *(shavasan)* calms the mind and relaxes the body.
- Dry hair immediately after washing to prevent sinus problems.
- Blowing the nose forcibly may be injurious to the ears, eyes and nose.
- Continuous nose picking and scratching the anus may be a sign of worms in the body.

- Long fingernails may be unhygienic.
- Cracking the joints may be injurious to the body (causes deranged *vata*).
- Repeated masturbation may be injurious to the body (causes derangement).
- It is harmful to have sex during menstruation (causes deranged *vata*).
- After sex, milk heated with raw cashews and raw sugar promotes strength and maintains sexual energy.
- Oral and anal sex are unhygienic (cause *vata* derangement).
- Sex immediately after meals is injurious to the body.
- Avoid physical exertion such as yoga or running during menstruation.

## Mental Hygiene

- Fear and nervousness dissipate energy and aggravate *vata*.
- Possessiveness, greed and attachment enhance *kapha*.
- Worry weakens the heart.
- Hate and anger create toxins in the body and aggravate *pitta*.
- Excessive talking dissipates energy and aggravates *vata*.

# Annotated Bibliography

*Note*: Citations in the text are in standard format: author's surname, page number, and date in parentheses ( ). A different system is used for Joseph Needham's *Science and Civilization in China*, because volume numbering and authorship on the actual volumes is stated in a confusing way. Accordingly, citations are in the following format: SCC volume number: section number page. Individual volumes and sections are listed in this bibliography under the series author Joseph Needham in order of date with the details of authorship given for each.

Adler, Joseph A., trans. 2002. Chu Hsi [Zhu Xi]. *Introduction of the Study of the Classic of Change* (I-hsueh ch'i-meng). Provo, UT: Global Scholarly Publications. Of interest because it gives a sense of the extreme complexity which Chinese numerology developed.

Austin, James H. 1998. *Zen and the Brain*. Cambridge, MA: MIT Press. Written by a neurologist who is an accomplished Zen Buddhist practitioner, this is one of very few books on the physiology of meditation that is medically sound.

Ayer, Alfred Jules. 1946. *Language, Truth and Logic*. London: Gollantz. Reprinted New York: Dover, n.d. The manifesto of the extreme positivist movement.

Aylward, Thomas E. 2007. *The Imperial Guide to Feng Shui and Chinese Astrology: The Only Authentic Translation from the Original Chinese*. London: Watkins Publishing.

Balkin, Jack M. 2002. *The Laws of Change: I Ching and the Philosophy of Life*. New York: Schoken Books. A Yale law professor presents the wisdom of the ancient Chinese classic from an entirely rationalistic viewpoint. Of interest as an example of how an ancient system can be transformed into one giving advice on modern life.

Barbour, Ian G. 1997. *Religion and Science: Historical and Contemporary Issues*. San Francisco: HarperSanFrancisco. A standard text cataloguing the various ways science and religion can be related to each other.

Barnes, Linda L. 2005. *Needles, Herbs, Gods and Ghosts: China, Healing and the West to 1848*. Cambridge, MA: Harvard University Press. The history of Chinese medicine as understood and practiced by Westerners. Shows convincingly that the current Western use of acupuncture is not new but has waxed and waned since the nineteenth century. Also interesting coverage of the disagreement between anatomy as revealed by dissection and the Chinese system of channels and virtual organs.

Basham, A. L. 1954. *The Wonder That Was India*. New Delhi, India: Rupa & Co. Reprinted 1981. Still one of the best introductions to traditional Indian culture.

Berger, Kenneth J. 1997. "Environment and Nature: China." In Helaine Selin, ed., *Encyclopedia of the History of Science, Technology, and Medicine in Non-Western Countries*. Dordrecht: Kluwer Academic, 291–293.

Bodde, Dirk. 1991. *Chinese Thought, Society and Science: The Intellectual and Social Background of Science and Technology in Pre-modern China*. Honolulu: University of Hawaii Press. Excellent overview of the social and historical factors affecting science in China with a more balanced discussion than that of Needham regarding the limitations of empirical method in China.

Bodhi, Bhikku. 2000. *A Comprehensive Manual of Abhidhamma: The Abhidhammattha Sangaha of Acarya Anuruddha*. Seattle, WA: BPS Pariyatti Editions. This very difficult work gives a feeling for how the early Buddhists analyzed mental phenomena.

Bresciani, Umberto. 2001. *Reinventing Confucianism: The New Confucian Movement*. Taipei: Taipei Ricci Institute for Chinese Studies. Articles on modern Chinese philosophers in the Confucian tradition. Discussion of science is scattered through the volume.

Brockington, John L. 1996. *The Sacred Thread*. New York: Columbia University Press. Bruun, Ole. 2003. *Fengshui in China: Geomantic Divination between State Orthodoxy and Popular Religion*. Honolulu: University of Hawai'i Press.

Bruun, Ole. 2003. *Fengshui in China: Geomantic Divination Between State Orthodoxy and Popular Religion*. Honolulu: University of Hawai'i Press.

Capra, Fritjof. 1991. *The Tao of Physics: An Exploration of the Parallels between Modern Physics and Eastern Mysticism*, 3rd ed., updated. Boston, MA: Shambhala. Revision of original 1975 edition. Entirely unreliable but unfortunately highly influential.

Chakravarty, A. K. 1997 "Calendars in India." In Helaine Selin, ed., *Encyclopedia of the History of Science, Technology, and Medicine in Non-Western Countries*. Dordrecht: Kluwer Academic, 168–171.

Chapple, Christopher Key, and Tucker, Mary Evelyn, eds. 2000. *Hinduism and Ecology: The Intersection of Earth, Sky, and Water*. Cambridge, MA: Harvard University Press. Reprinted 2001, Delhi, India: Oxford University Press. Modern ecological interpretations of Hindu texts. Little on actual human-environmental interaction in traditional India.

Chattopadhiyaya, D. P. 1997. "Environment and Nature: India." In Helaine Selin, ed., *Encyclopedia of the History of Science, Technology, and Medicine in Non-Western Countries*. Dordrecht: Kluwer Academic, 295f.

Chen, Cheng-Yih. 1996. *Early Chinese Work in Natural Science: A Re-examination of the Physics of Motion, Acoustics, Astronomy and Scientific Thoughts.* Hong Kong: Hong Kong University Press. Technical coverage of the mathematically based sciences in China for those with a background in physics. Gives a different perspective on the nature of Chinese science.

Cheng, Chung-yin. 2003. "*Qi (Ch'i):* Vital Force." In Anthony S. Cua, ed., *Encyclopedia of Chinese Philosophy.* New York and London: Routledge, 615–617.

Cleary, Thomas, trans. 1993. *The Flower Ornament Scripture.* Boston, MA: Shambhala Publications. The most elaborate of the cosmological Mahayana Buddhist sutras. Gives a sense of the cosmos as imagined by this Buddhist tradition.

Cremo, Michael A. 1998. *Forbidden Archeology.* Badger, CA: Torchlight Publishing. A pseudoscientific attempt to make archeology fit the Vedic chronology. An example of the fallacy that science can be adduced to prove religion.

Cua, Anthony S., ed. 2003. *Encyclopedia of Chinese Philosophy.* New York: Routledge.

Danielou, Alain. 1964. *The Gods of India: Hindu Polytheism.* Trans. from the French. Reprinted 1985, New York: Inner Traditions International. Excerpts from Hindu sacred texts organized to present concepts clearly. Not always reliable but a work of literary elegance.

Das, Rahul Peter. 2003. *The Origin of the Life of a Human Being: Conception and the Female According to Ancient Indian Medical and Sexological Literature.* Delhi, India: Motilal Banarsidass. Very detailed study of sexuality and reproduction in traditional Indian texts.

Dash, Bhagwan. 1997. "Sasruta." In Helaine Selin, ed., *Encyclopedia of the History of Science, Technology, and Medicine in Non-Western Countries.* Dordrecht: Kluwer Academic, 927.

Desai, Prakash N. 1997. "Medical Ethics in India." In Helaine Selin, ed., *Encyclopedia of the History of Science, Technology, and Medicine in Non-Western Countries.* Dordrecht: Kluwer Academic, 669–671.

Dyson, Freeman J. March 28, 2002. "Science and Religion: No Ends in Sight." *New York Review of Books.*

Eckman, Peter. 1996. *In the Footsteps of the Yellow Emperor: Tracing the History of Traditional Acupuncture.* San Francisco, CA: Cypress Book Company.

Elman, Benjamin A. 2001. *From Philosophy to Philology: Intellectual and Social Aspects of Change in Late Imperial China,* 2nd rev. ed. Los Angeles: UCLA Asian Pacific Monograph Series.

———. 2005. *On Their Own Terms: Science in China, 1550–1900.* Cambridge, MA: Harvard University Press. Covers the final phases of traditional Chinese sciences and the early interactions with Western science.

———. 2006. *A Cultural History of Modern Science in China.* Cambridge, MA: Harvard University Press. Describes the final stages in the adaptation of Western science in China with some coverage of adaptation of Chinese technology by Europe.

Elvin, Mark. 2004. *The Retreat of the Elephants: An Environmental History of China.* New Haven, CT: Yale University Press. The definitive work on ecology in China. Much material available nowhere else in English on actual effects of Chinese society on its natural surroundings. Does not confuse philosophical ideals with actual behavior. Should be read by anyone interested in the history of human interaction with the environment.

Erbaugh, Mary S., ed. 2002. *Difficult Characters*. Columbus: Ohio State University National East Asian Language Resource Center.

Farquhar, Judith. 1994. *Knowing Practice: The Clinical Encounter of Chinese Medicine*. Boulder, CO: Westview Press. Written by a medical anthropologist, this work gives a much better sense of how traditional medicine is actually practiced in China. Gives a good sense of the thought processes of practitioners of TCM and hence provides a dimension missing in translations of textbooks.

Feng, Gia-fu, and English, Jane. 1972. *Lao Tsu Tao Te Ching: A New Translation*. New York: Vintage. The most readable translation of this great work of world philosophical literature. Closer to the spirit of the original than most scholarly translations.

Feuchtwang, Stephen. 2002. *An Anthropological Analysis of Chinese Geomancy*. Bangkok, Thailand: White Lotus.

Fields, Gregory P. 2001. *Religious Therapeutics: Body and Health in Yoga, Ayurveda, and Tantra*. Albany: State University of New York Press. Covers the relation of body, mind, and spirit in traditional Indian thought to concepts of healing.

Filliozat, Jean. 1964. *The Classical Doctrine of Indian Medicine: Its Origins and Its Greek Parallels*. New Delhi, India: Munshiram Manoharlal. Still a useful work by one of the first Western scholars on Indian medicine.

————. 1991. *Religion, Philosophy, Yoga*. Delhi, India: Motilal Banarsidass. Filliozat was an ophthalmologist who became Professor of Indology at the Ecole Pratique des Hautes Etudes in Paris. Because of his medical background, much of his writing concerns the relation of science to spirituality. This makes his essay in this volume, "The Limits of Human Powers in India" one of the best sources on the extreme physical capabilities of yogins.

Furth, Charlotte, Zeitlin, Judith T., and Hsiung, Ping-chen. 2007. *Thinking with Cases: Specialist Knowledge in Chinese Cultural History*. Honolulu: University of Hawai'i Press.

Ganeri, Jonardon. 2001. *Philosophy in Classical India: The Proper Work of Reason*. London: Routledge. Useful introduction.

Garfield, Jay L. 1995. *The Fundamental Wisdom of the Middle Way: Nagarjuna's Mulamadhyamakakarika*. Oxford: Oxford University Press. One of the least confusing works on the great Buddhist logician, Nagarjuna. Of interest in presenting a mode of reasoning quite different from that of modern analytic philosophy.

Girardot, N. J., Miller, James, and Liu Xiaogan., eds. 2001. *Daoism and Ecology: Ways Within a Cosmic Landscape*. Cambridge: Center for the Study of World Religions and Harvard University Press.

Goldin, Paul Rakita. 1999. *Rituals of the Way: The Philosophy of Xunzi*. Chicago: Open Court/Carus.

Hadot, Pierre. 2002. *What Is Ancient Philosophy?* Cambridge, MA: Belknap Press of Harvard University Press.

Hall, Manly P. 1995. *Astrological Keywords: Compiled from Leading Authorities.* Los Angeles: Philosophical Research Society.

Hanson, Chad. 1992. *A Daoist Theory of Chinese Thought: A Philosophical Interpretation.* Oxford: Oxford University Press. An attempt to interpret early Chinese philosophy as if it wanted to be Western philosophy.

Harbsmeier, Christoph. 1995. "Some Notions of Time and History in China and in the West." In Chun-chieh Huang and Erik Zurcher, eds., *Time and Space in Chinese Culture.* Leiden: E. J. Brill, 49–71.

Hayden, Brian. 2003. *Shamans, Sorcerers and Saints: A Prehistory of Religion.* Washington, DC: Smithsonian Books.

Henderson, John B. 1984. *The Development and Decline of Chinese Cosmology.* New York: Columbia University Press. The standard work on Chinese cosmology and invaluable for study of the Chinese worldview.

———. 2003a. "Cosmology." In Antonio S. Cua, ed., *Encyclopedia of Chinese Philosophy.* New York: Routledge, 187–194.

———. 2003b. "Wuxing (Wu-hsing): Five Phases." In Antonio S. Cua, ed., *Encyclopedia of Chinese Philosophy.* New York: Routledge, 786–788.

Ho Peng Yoke. 2003. *Chinese Mathematical Astrology: Reaching to the Stars.* London: RoutledgeCurzon. Quite technical but gives a fascinating picture of the immense mathematical complexity of Chinese astrology.

Holden, James Herschel. 1996. *A History of Horoscopic Astrology: From the Babylonian Period to the Modern Age.* Tempe, AZ: American Federation of Astrologers.

Hon, Tze-ki. 2003. "Cheng Yi (Ch'eng I)." In Anthony S. Cua, ed., *Encyclopedia of Chinese Philosophy.* New York: Routledge, 43–46.

———. 2005. *The Yijing and Chinese Politics: Classical Commentary and Literati Activism in the Northern Song Period 960–1127.* Albany: State University of New York Press.

Hong, Yingming, White, Paul, and Jiang, Frank, trans. 2003. *Tending the Roots of Wisdom.* Beijing, China: New World Press.

Huff, Toby E. 1993. *The Rise of Early Modern Science: Islam, China and the West.* Cambridge: Cambridge University Press. Attributes the incomplete development of science in China and Islam to lack of an institutional structure comparable to the universities of Europe, which provided for knowledge to be preserved, developed, and transmitted by a pluralistic faculty. One of the most important works on this subject.

Jaspers, Karl. 1932. *Philosophie.* Translated by E. B. Ashton as *Philosophy,* Vol. I. Chicago: University of Chicago Press, 1969.

Ji, Fengyuan. 2004. *Linguistic Engineering: Language and Politics in Mao's China.* Honolulu: University of Hawai'i Press.

Keightley, David N., ed. 1983. *The Origins of Chinese Civilization.* Berkeley: University of California Press.

Kim, Yung Sik. 2000. *The Natural Philosophy of Chu Hsi 1130–1200.* Philadelphia, PA: American Philosophical Society. Shows that Zhuxi, one of China's greatest philosophers, was interested in science as well as philosophy. Kim is a scientist himself and does an outstanding job of bringing out this aspect of Zhu's philosophy that has tended to be neglected.

Knoblock, John, and Riegel, Jeffrey. 2000. *The Annals of Lu Buwei: A Complete Translation and Study*. Stanford, CA: Stanford University Press. A translation of this important Qin dynasty text which attempted to record all important knowledge that existed at the time. Invaluable for understanding Chinese thought at the time China was transformed from a collection of small states to a unified empire. Lu was a tragic figure, having attained high rank, he lost out in court intrigues and committed suicide to avoid execution.

Koh, Vincent. 2003. *Basic Science of Feng Shui: A Handbook for Practitioners*. Singapore: Asiapac Books PTE. Most English language books on feng shui are really just about home decoration spiced with small bits of feng shui lore. Koh's book presents his school of feng shui in full detail. Not historical but a good place to see how feng shui is still being done in Asia.

Kuhn, Thomas. 1970. *The Structure of Scientific Revolutions*, 2nd ed., enlarged. Chicago: University of Chicago Press.

Kunst, Richard. 1985. *The Original "Yijing": A Text, Phonetic Description, Translation, and Indexes, with Sample Glosses*. Doctoral dissertation. Ann Arbor, MI: University Microfilms International.

Lad, Vasant. 1984. *Ayurveda, The Science of Self-Healing: A Practical Guide*. Wilmot, WI: Lotus Press. Written by a prominent modern practitioner. Not historical but a lucid guide to the basics of traditional Indian medicine.

Lau, D. C. 2003. *Mencius: A Bilingual Edition*. Hong Kong: Chinese University Press.

Le Blanc, Charles. 1993. "Huai nan tzu." In Michael Loewe, ed., *Early Chinese Texts: A Bibliographic Guide*. Berkeley: Society for the Study of Early China and Institute of East Asian Studies, University of California.

Legge, James, trans. 1892. *The Chinese Classics, Vol. I: Confucian Analects, The Great Learning, The Doctrine of the Mean*. Taipei, Republic of China: SMC Publishing. Reprinted 1998. Legge's translations are the earliest English versions of the Chinese classics with any degree of accuracy but are rather stilted. They exist in many reprint editions.

Legge, James, trans. 1895. *The Chinese Classics, Vol. II: The Works of Mencius*. Taipei, Republic of China: SMC Publishing. Reprinted 1998.

———. 1899. *The Sacred Books of China: The I Ching*, 2nd ed. New York: Dover. Reprinted 1963.

Lewis, David. 1986. *On the Plurality of Worlds*. Oxford, U.K.: Basil Blackwell.

Little, Stephen with Eichman, Shawn. 2000. *Taoism and the Arts of China*. Chicago and Berkeley: Art Institute of Chicago and University of California Press. The only scholarly work in English on the visual culture of Daoism.

Liu, Fang-ju, and Brix, Donald E., trans. 1997. *Blessings for the New Year: Special Exhibition of Paintings of Chung K'uei*. Taipei, China: The National Palace Museum.

Lloyd, Geoffrey, and Sivin, Nathan. 2002. *The Way and the Word: Science and Medicine in Early China and Greece*. New Haven, CT: Yale University Press.

Lowe, Michael. 1982. *Chinese Ideas of Life and Death: Faith, Myth and Reason in the Han Period (202 BC–AD 220)*. Reprinted 1994 Taipei, China: SMC Publishing.

————, ed. 1993. *Early Chinese Texts: A Bibliographical Guide*. Berkeley, CA: Society for the Study of Early China and Institute of East Asian Studies, University of California.

Lynn, Richard John, trans. 1994. *The Classic of Changes: A New Translation of the I Ching as Interpreted by Wang Bi*. New York: Columbia University Press. Generally considered the most accurate translation of this classic but somewhat ponderous.

Mair, Victor H., Steinhardt, Nancy S., and Goldin, Paul R., eds. 2005. *Hawai'i Reader in Traditional Chinese Culture*. Honolulu: University of Hawai'i Press. Excellent anthology with a wealth of material on all aspects of Chinese culture, including both science and religion.

Major, John S. 1993. *Heaven and Earth in Early Han Thought: Chapters Three, Four and Five of the Huainanzi*. Albany: State University of New York Press.

McEvilley, Thomas. 2002. *The Shape of Ancient Thought: Comparative Studies in Greek and Indian Philosophies*. New York: Allworth Press and School of Visual Arts. Striking similarities have long been noted between the philosophies of India and those of the Ancient Near East and Greece. This is the definite work on this subject although the underlying issue of how much these similarities are due to contact and how much independent origin is likely to remain unresolved.

McKnight, Brian E., trans. 1981. Sung Ts'u *The Washing Away of Wrong. Science, Medicine, &Technology in East Asia 1*. Ann Arbor, MI: Center for Chinese Studies, The University of Michigan. An early text on forensic pathology notable for full development of empirical method.

Needham, Joseph. 1981. *Science in Traditional China: A Comparative Perspective*. Cambridge, MA: Harvard University Press, and Hong Kong: The Chinese University Press.

Needham, Joseph. 1954 (and subsequent). *Science and Civilization in China*. Cambridge: Cambridge University Press. A unique work of scholarship that is both authoritative and elegantly written.

Needham, Joseph with the collaboration of Wang Ling. 1959. *Mathematics and the Science of the Heavens and the Earth*, Vol. 3.

————. 1998. Volume VII:1 by Christoph Harbsmeier: *Language and Logic*.

————. 2000. Volume VII:6 *Biology and Biological Technology: Medicine*. By Joseph Needham with the collaboration of Lu Gwei Djen. Edited with an Introduction by Nathan Sivin.

————. 2004. Volume VII:2 *General Conclusions and Reflections*. by Joseph Needham, with the collaboration of Kenneth Girdwood Johnson and Ray Huang (Huang Jen-yu).

————. 2004. Volume V:12 *Ceramic Technology* Rose Kerr and Nigel Wood. The definitive work on Chinese ceramic technology with some coverage of the patron gods of potters and of cultural aspects. Does not cover style and aesthetics in any detail.

Olson, Richard. 2004. *Science and Religion, 1450–1900: From Copernicus to Darwin*. Westport, CT: Greenwood Press.

Pankenier, David. 2005. "Astronomy in Early Chinese Sources." in Victor H. Mair, Nancy S. Steinhardt, and Paul R. Goldin, eds., *Hawai'i Reader in Traditional Chinese Culture*. Honolulu: University of Hawai'i Press, 18–27.

Poo, Mu-Chou. 1990. "Ideas Concerning Death and Burial in Pre-Han and Han China." *Asia Major* (Third Series), Vol. III, Part 2, 25–62.

Principe, Lawrence W., and Newman, William R. 2001. "Some Problems with the Historiography of Alchemy." In William R. Newman and Anthony Grafton, eds., *Secrets of Nature: Astrology and Alchemy in Early Modern Europe*. Cambridge, MA: The MIT Press.

Rahman, A. 1998. "A Perspective of Indian Science." In A. Rahman, ed., *History of Indian Science, Technology and Culture AD 1000–1800*. New Delhi, India: Oxford University Press

———. 2002. "India's Interaction with China, Central and West Asia." Vol. III, Part 2 of *History of Science, Philosophy and Culture in Indian Civilization*. General ed. D. P. Chattopadhyaya. New Delhi, India: Oxford University Press.

Redmond, Geoffrey. 1995. "Eugenics and Religious Law: Hinduism and Buddhism." In W. T. Reich, ed., *Encyclopedia of Bioethics*, rev. ed. New York: Macmillan, 784–788.

———, ed. 1996. *Comparing Buddhism and Science. The Pacific World Journal of the Institute of Buddhist Studies* (New Series). Special issue on Buddhism: Medicine Science and Technology

———. 2004. Review of Allan Hunt Badiner and Alex Grey: *Zig Zag Zen: Buddhism and Psychedelics*. San Francisco: Chronicle Books 2002. *Journal of Buddhist Ethics*.

Rochat de la Vallee, Elisabeth. in Instituts Ricci, Desclee de Bouwer. 2003. *Apercus de Civilisation Chinoise: Les Dossiers du Grand Ricci*. Paris: Association Ricci – Desclee de Brouwer. Extensive tabular material on many aspects of Chinese civilization including cosmology and science. Excerpted from the major Western language Chinese dictionary. Accessible with limited French.

Ronkin, Noa. 2005. *Early Buddhist Metaphysics: The Making of a Philosophical Tradition*. London: RoutledgeCurzon.

Rutt, Richard. 1996. *The Book of Changes (Zhouyi): A Bronze Age Document*. Richmond, Surrey: Curzon Press. An attempt to recover the original meaning of this maddeningly obscure text. Should be compared to the translations of Lynn and Wilhelm to see how greatly scholarly interpretations can diverge. Excellent introduction with background material.

Sadakata, Akira. 1997. *Buddhist Cosmology: Philosophy and Origins*. Tokyo, Japan: Kosei Publishing Co.

Samian, Abdul Latif. 1997. "Al-Biruni (Part 1)." In Helaine Selin, ed., *Encyclopedia of the History of Science, Technology, and Medicine in Non-Western Countries*. Dordrecht: Kluwer Academic.

Sarma, K. V. 1997. "Astronomy in India." In Helaine Selin, ed., *Encyclopedia of the History of Science, Technology, and Medicine in Non-Western Countries*. Dordrecht: Kluwer Academic.

Schafer, Edward H. 2005. *Pacing the Void: Tang Approaches to the Stars*. Berkeley: University of California Press. 1977 Reprinted Floating World Editions . Entertaining but rigorously scholarly account of Chinese fantasy journeys to remote realms. These Chinese accounts are not unlike modern science fiction.

Schwab, Raymond. 1984. *The Oriental Renaissance: Europe's Rediscovery of India and the East, 1680–1880*. Translated by Gene Patterson-Black and Victor Reinking. New York: Columbia University Press (Original in French, 1950). Traces the origins of the philology of Asian languages, particularly Sanskrit, in Europe.

Selin, Helaine, ed. 1997. *Encyclopedia of the History of Science, Technology, and Medicine in Non-Western Countries*. Dordrecht: Kluwer Academic. A very valuable reference source providing difficult to find information. As with most multi-authored works accuracy is uneven. The sections on China are more reliable than those on India.

Shaughnessy, E. L. 1983. The Composition of the "Zhouyi." Doctoral dissertation, Stanford University. Ann Arbor, MI: University Microfilms International. Important demonstration of astronomical references in the *Yi Jing*.

Sivin, Nathan. 1987. *Traditional Medicine in Contemporary China: A Partial Translation of* Revised Outline of Chinese Medicine (1972) *with an Introductory Study on Change in Present Day and Early Medicine*. Ann Arbor, MI: Center for Chinese Studies, University of Michigan. A manual from the Maoist era. Notable for making careful translations of Chinese medical terminology.

———. 1995. *Medicine, Philosophy and Religion in Ancient China*. Aldershot, Hampshire: Variorum/Ashgate. Collected papers by the leading American scholar of Chinese science and medicine and essential reading for any interested in these subjects. Especially valuable for its critique of common errors in modern scholarship.

Smith, Brian K. 1994. *Classifying the Universe: The Ancient Indian* Varna *System and the Origins of Caste*. Oxford: Oxford University Press. Definitive work on ancient Indian correlative cosmology with emphasis on the caste system.

Smith, Kidder, Jr., Bol, Peter K., Adler, Joseph, and Wyatt, Don J. 1990. *Sung Dynasty Uses of the I Ching*. Princeton, NJ: Princeton University Press.

Soka Gakkai. 2002. *The Soka Gakkai Dictionary of Buddhism*. Tokyo, Japan: Soka Gakkai. A concise source for much useful information about Buddhism though emphasis is on the aspects related to the Japanese Soka Gakkai sect, which is based on the Lotus Sutra.

Strickmann, Michel. 2002. *Chinese Magical Medicine*. Edited by Bernard Faure. Stanford, CA: Stanford University Press. Most Western works on Chinese medicine have emphasized the rational system based on yin-yang, qi, and wu xing. This covers some of the more religious healing practices.

Sung, E. 1999. *Ten Thousand Years Book*. San Francisco, CA: MJE Publishing. Provides tables used to calculate Chinese four pillar horoscopes.

Temple, Robert. 1998. *The Genius of China: 3,000 Years of Science, Discovery and Invention*. London: Prion Books. A popularization of Needham's work. Valuable

for its excellent illustrations but many of the text's claims of Chinese discoveries preceding Western ones are dubious. The text is best bypassed in
favor of Needham's own work.

Terzani, Tiziano. 1997. *A Fortune Teller Told Me: Earthbound Travels in the Far East.*
New York: Three Rivers Press. The author, an Italian journalist long resident in Asia, spent a year traveling to visit fortune-tellers. While not an
anthropological account, it gives a vivid account of the place of divination
in the life of modern Asia and so is a valuable supplement to scholarly
studies based on analysis of texts.

Tremlin, Todd. 2006. *Minds and Gods: The Cognitive Foundations of Religion.* Oxford:
Oxford University Press.

Unschuld, Paul. 1985. *Medicine in China: A History of Ideas.* Berkeley: University
of California Press. The single best work on the subject and the place to
start for anyone with a serious interest. Unschuld's breadth of knowledge
is extraordinary. His writing is lucid and he brings critical scholarship to a
field greatly in need of it.

———. 1986. *Medicine in China: A History of Pharmaceutics.* Berkeley: University of
California Press.

———. 1998. *Forgotten Traditions of Ancient Chinese Medicine: A Chinese View from
the Eighteenth Century. The I-hsueh Yuan Liu Lun of 1757.* Brookline, MA:
Paradigm Publications. A series of essays on medical topics by a mid-Qing
physician. Because it is written in a personal style it gives a sense of how
Chinese doctors actually thought about what they did.

———. 2003. *Huang Di Nei Su Wen: Nature, Knowledge, Imagery in an Ancient Chinese
Medical Text.* Berkeley: University of California Press. The definitive work
on the Su Wen, usually mistranslated as "The Yellow Emperor's Classic of
Internal Medicine."

Victoria, Brian (Daizen) A. 1997. *Zen at War.* New York: Weatherhill.

Volkmar, Barbara. 2004. "On Sense and Non-Sense of Premodern Medical Theories:
The Example of Theories on Smallpox." In *Discussions of Ancient and Modern.*
Taiwan: Academia Sinica. Excellent coverage of smallpox prevention in
China.

Walters, Derek. 1987. *The Complete Guide to Chinese Astrology.* London: Watkins.
Reprinted 2002. Much interesting information though written for potential
believers.

———. 1991. *The Fengshui Handbook: A Practical Guide to Chinese Geomancy and
Environmental Harmony.* London: Thorsons.

Watson, Burton, trans. 1968. *The Complete Works of Chuang Tzu.* New York:
Columbia University Press. Zhuangzi is unique in the history of philosophy; this is a classic of world literature.

White, David Gordon. 1996. *The Alchemical Body: Siddha Traditions in Medieval India.*
Chicago: University of Chicago Press. Detailed but lucid treatment of the
intricacies of Indian body-based spirituality.

Wilhelm, Richard. 1967. *The I Ching or Book of Changes.* The Richard Wilhelm Translation rendered into English by Cary F. Baynes. Foreword by C. G. Jung.
Preface to the Third Edition by Hellmut Wilhelm. Princeton, NJ: Princeton

University Press. Bollingen Series XIX The translation that made the Yi Jing a classic in the West. Basically sound.Though it rearranges sections of the original in an eccentric way, this translation based on the classic as viewed by Chinese literati in the early twentieth century brings out the appeal of this work as no other translation has.

Wilkinson, Endymion. 1998. *Chinese History: A Manual*. Cambridge, MA: Harvard University Asia Center.

Wittgenstein, Ludwig. 1993. "Remarks on Frazer's *Golden Bough*." In James Klagge and Alfred Nordmann, eds., *Philosophical Occasions 1912–1951*. Indianapolis, IN: Hackett Publishing Company.

Wu, Jing-Nuan. 1993. *Lingshu or The Spiritual Pivot*. Washington, DC: The Taoist Center, and Honolulu: University of Hawai'i Press. A translation of the early classic on acupuncture.

Xu, Gan. 2002. *Balanced Discourses: A Bilingual Edition*. New Haven, CT: Yale University Press, and Beijing, China: Foreign Languages Press.

Yang, Yifang. 2002. *Chinese Herbal Medicines: Comparisons and Characteristics*. London: Churchill Livingstone. An example of a text for use by modern practitioners.

Yang Xin, Nie Chongzheng, Lang Shaojun, Barnhart, Richard M., Cahill, James, and Wu Hung. 1997. *Three Thousand Years of Chinese Painting*. New Haven, CT: Yale University Press. An excellent work for familiarizing oneself with the great painting tradition of China. The depiction of landscapes and animals reveals the Chinese feeling for nature.

Yao, XinZhong. 2003. *RoutledgeCurzon Encyclopedia of Confucianism*. London: RoutledgeCurzon.

Zukav, Gary. 1979. *The Dancing Wu Li Masters*. New York: Bantam. Reprinted 1980. Comparable to Capra's work in its New Age distortions of both science and Asian religion.

Zysk, Kenneth. 1991. *Asceticism and Healing in Ancient India: Medicine in the Buddhist Monastery*. Oxford: Oxford University Press. The best concise treatment of early Indian medicine, outstanding for employing a critical historical approach in a field that too often fails to distinguish myth from fact.

———. 1993. *Religious Medicine: The History and Evolution of Indian Medicine*. New Brunswick, NJ: Transaction.

# Index

Chinese chronology, xvii–xix; Former
(Western) Han (206 B.C.E.–8 C.E.)
and Latter (Eastern) Han (25– 220
C.E.), xviii; Ming (1368–1644), xix;
Mythical times and Neolithic, xvii;
Northern Wei (386–584), xviii;
People's Republic of China (1949 to
present), xix; Qin (221–206 B.C.E.),
xviii; Qing (1636–1912), xix; Republic
(1912 to present. Continues only in
Taiwan), xix; Shang (1600–1045
B.C.E.), xvii; Song (960–1279),
xviii–xix; Tang (618–907), xviii;
Warring States (475–221 B.C.E.),
xviii; Western Zhou, xviii; Yuan
(1271–1368), xix
Chinese language, 7, 73–74; logic and,
75–78
Chinese written language, 44–45
Chola Empire (848–1279), xxi
Christianity, cosmology in, 55–56
Civil service examinations, in China,
54, 71, 196
*Classic of Change (Yi Jing)*, 45–50
*Classifying the Universe: The Ancient
Indian Varna System and the Origins of
Caste* (Smith), 217–18
Clepsydra (water clock), 98, 109
Colonial Period (1757–1947), xxii
Comets, 116
Comparative methodology, limitations
to, 4–5
Confucian texts: by Menicus, 52–53, 54;
*Lunyu*, 41–42, 46, 50–51, 54, 60, 102,
129; by Xunzi, 53; *Yi Jing (Classic of
Change)*, 45–50; *Zhouli*, 203–4; by
Zhuxi, 53, 54
Confucianism, 36; and astrology,
115–17; and astronomy/astrology,
115–17; bodily contact as taboo in,
161, 176–77; as closed system, 68;
consumption of animals, 52–53;
correction of improper behavior in,
51–52, 167; cosmology in, 48; Dao
(Way) in, 39; Daoism and Buddhism
as threat to, 43; human evil in, 53; on

knowledge, 75; on male–female
affection, 79; Neoconfucian
synthesis, 53–54; *wu tun* (five human
relationships), 26, 51; *zgeng ming*
(rectification of names), 51
Confucius: historical, 41–42, 52; on
music, 66; on social order, 102; on
spirits, 36, 68, 86, 140; on time, 97
Conspicuous consumption, in ancient
world, 192
Content *vs.* method, 16–17
Correlative anatomy and physiology,
159–61
Correlative classification, into groups
of five, 65–66
Correlative systems, 52, 61, 62;
inconsistencies in, 83
Correlative thought, 7–8, 17
Cosmology: defining, 55; scientific, 48,
55, 56–57; traditional, as metaphysic,
56. *See also* Metaphysics/cosmology,
in traditional China;
Metaphysics/cosmology, in
traditional India
Creationism, 57
Cultural context, and religion, 33

*Da Zhuan (Great Commentary)*, 46–47,
49, 205
Dalai Lama, 91, 199
Danielou, Alain, 218–22
Dao (Way), 39; multiple meanings of,
59–60; one and the many problem
and, 67. *See also Dao De Jing*; Daoism
*Dao De Jing*, 36, 59, 68; ecological
awareness and, 129, 130; logic in, 78;
on knowledge, 75; permanence and
change in, 102
Daoism: as threat to Confucism, 43;
astronomy/astrology in, 113–15;
cosmology in, 48; demonic healing
in, 154–55; dietary concepts, 159;
freedom of human imagination in,
113–15; on going against nature, 127,
128; meditation and internal organs
in, 177; meditation in, 43–44;

## About the Author

GEOFFREY REDMOND is the President of the Center for Health Research, Inc. and Director of the Hormone Center of New York. He is an expert on Chinese science, and the author of *Science and Eastern Religion: A Critical Reappraisal*. He is Director of Independent Scholars of Asia.